Clustering Dynamics and the Location of High-Tech-Firms

Contributions to Economics

http://www.springer.de/cgi-bin/search_book.pl?series=1262

Michael Carlberg
Intertemporal Macroeconomics
1998. ISBN 3-7908-1096-7

Sabine Spangenberg
The Institutionalised Transformation of the East German Economy
1998. ISBN 3-7908-1103-3

Hagen Bobzin
Indivisibilities
1998. ISBN 3-7908-1123-8

Helmut Wagner (Ed.)
Current Issues in Monetary Economics
1998. ISBN 3-7908-1127-0

Peter Michaelis/Frank Stähler (Eds.)
Recent Policy Issues in Environmental and Resource Economics
1998. ISBN 3-7908-1137-8

Jessica de Wolff
The Political Economy of Fiscal Decisions
1998. ISBN 3-7908-1130-0

Georg Bol/Gholamreza Nakhaeizadeh/
Karl-Heinz Vollmer (Eds.)
Risk Measurements, Econometrics and Neural Networks
1998. ISBN 3-7908-1152-1

Joachim Winter
Investment and Exit Decisions at the Plant Level
1998. ISBN 3-7908-1154-8

Bernd Meyer
Intertemporal Asset Pricing
1999. ISBN 3-7908-1159-9

Uwe Walz
Dynamics of Regional Integration
1999. ISBN 3-7908-1185-8

Michael Carlberg
European Monetary Union
1999. ISBN 3-7908-1191-2

Giovanni Galizzi/
Luciano Venturini (Eds.)
Vertical Relationships and Coordination in the Food System
1999. ISBN 3-7908-1192-0

Gustav A. Horn/Wolfgang Scheremet/
Rudolf Zwiener
Wages and the Euro
1999. ISBN 3-7908-1199-8

Dirk Willer
The Development of Equity Capital Markets in Transition Economies
1999. ISBN 3-7908-1198-X

Karl Matthias Weber
Innovation Diffusion and Political Control of Energy Technologies
1999. ISBN 3-7908-1205-6

Heike Link et al.
The Costs of Road Infrastructure and Congestion in Europe
1999. ISBN 3-7908-1201-3

Simon Duindam
Military Conscription
1999. ISBN 3-7908-1203-X

Bruno Jeitziner
Political Economy of the Swiss National Bank
1999. ISBN 3-7908-1209-9

Irene Ring et al. (Eds.)
Regional Sustainability
1999. ISBN 3-7908-1233-1

Katharina Müller/Andreas Ryll/
Hans-Jürgen Wagener (Eds.)
Transformation of Social Security: Pensions in Central-Eastern Europe
1999. ISBN 3-7908-1210-2

Stefan Traub
Framing Effects in Taxation
1999. ISBN 3-7908-1240-4

Pablo Coto-Millán
Utility and Production
1999. ISBN 3-7908-1153-X

Frank Riedel
Imperfect Information and Investor Heterogeneity in the Bond Market
2000. ISBN 3-7908-1247-1

Kirsten Ralf
Business Cycles
2000. ISBN 3-7908-1245-5

Michele Bagella/
Leonardo Becchetti (Eds.)
The Competitive Advantage of Industrial Districts
2000. ISBN 3-7908-1254-4

continued on page 245

Mario A. Maggioni

Clustering Dynamics and the Location of High-Tech-Firms

With 44 Figures and 33 Tables

Physica-Verlag
A Springer-Verlag Company

Series Editors
Werner A. Müller
Martina Bihn

Author
Professor Mario A. Maggioni
DISEIS
Università Cattolica Milano
Largo Gemelli, 1
20123 Milano
Italy
maggioni@mi.unicatt.it

ISSN 1431-1933
ISBN 3-7908-1431-8 Physica-Verlag Heidelberg New York

Cataloging-in-Publication Data applied for
Die Deutsche Bibliothek – CIP-Einheitsaufnahme
Maggioni, Mario A.: Clustering dynamics and the location of high tech firms: with 33 tables / Mario A. Maggioni. – Heidelberg; New York: Physica-Verl., 2002
(Contributions to economics)
ISBN 3-7908-1431-8

Physica-Verlag Heidelberg New York
a member of BertelsmannSpringer Science+Business Media GmbH

Softcover Design: Erich Kirchner, Heidelberg

SPIN 10848874 88/2202-5 4 3 2 1 0 – Printed on acid-free and non-aging paper

To Nani, Marta, Anna, Pinuccia and Tarcisio

How did Warsaw become Warsaw? First they built one house, then another, and gradually a city emerged. Everything grows. Even stones grow.
I.B. Singer (1962), *Stories for Children.*

Meytl, meytl, ch'vel bay dir fregn:
vos ken vaksn, vaksn on regn? (...)
Narisher bokher, vos darfstu fregn?
A shteyn ken vaksn, vaksn on regn
Tum balalayka, (popular Yiddish song)

Maiden, maiden I'd like to ask you:
what can grow without rain? (...)
Silly lad, why do you ask?
A stone can grow without rain
(English translation by Klezmer Conservatory Band)

Unless the Lord builds the house, its builders labour in vain.
Psalm 127

Acknowledgements

I would like to thank many people who have, in different ways, helped and encouraged me in the process of writing this thesis.

First of all I would like to thank my supervisor, Paul Stoneman, for his valuable comments and suggestions and for the patience he has shown with me. Writing a thesis while working is a long, painful experience and without Paul's guidance I would have given up the task before its completion.

I would like to thank Alberto Quadrio Curzio who first introduced me to the study of economics (conceived as a discipline composed of pure theory and applied analysis) and has always encouraged my studies and research. I also want to express my gratitude to Gilberto Antonelli for his guidance and friendship during my stay at IDSE and beyond. I would also like to thank my past colleagues at IDSE-CNR and my current colleagues at the Catholic University for having made my journeys to Warwick possible.

During these years I have benefited from comments and suggestions from G. Antonelli, G. Battisti, M. Baussola, C. Beretta, S. Beretta, A. Bramanti, E. Ciciotti, G. Folloni, E. Giacobino, F. Gambarotto, T. Leinbach, C. Longhi, C. Lucifora, M. Luger, G. Merzoni, J.S. Metcalfe, M. Nosvelli, G. Porro, P. Quintas, A. Reggiani, L. Senn, M. Syrquin, M. Storper, E. Uberti and two anonymous referees of *Regional Studies*.

I would also like to thank, for their comments, the participants at the XXII EARIE (European Association for Research in Industrial Economics) Conference, Juan Les Pins, 1995; XXV RSAI (Regional Science Association International, English and Irish Section) Conference, Dublin, 1994; XV RSAI (Regional Science Association International, North American Section) Conference, Houston,1993; XV AISRe (Regional Science Association International, Italian Section) Conference, Bologna 1993, and at the seminars held at Bocconi University, Milan, 1998; IDSE, Milan, 1997 and 1995; Latapses, Sophia-Antipolis, 1997; University of Siena, Siena, 1996; University of Trento, Trento, 1995.

C. Longhi (Latapses-CNRS, France), Y.D. Funderburk and M. Italiano (Census Bureau, USA), S. Mackey (National Conference of State Legislatures, USA), C. Powell (Office of National Statistics, UK) have been very helpful in the process of data collection. M. Nosvelli helped me with GIS data mapping, F. Canal and L. Maggioni rescued me when my computer crashed in France and in the UK.

I would like to gratefully mention the institutions which have supported my studies and research; in particular IDSE-CNR, where I spent a large and important part of my career, and CRANEC (Centre for Economic Analysis Research at the Catholic University of Milan) where I started my research and am currently working. Financial support from CRANEC, IDSE-CNR, ISEIS, CNR "Short Term

Mobility Programme", and CNR-CNRS "bilateral exchange programmes" is also gratefully acknowledged. Finally Murst research programme *Infrastrutture, competitività, livelli di governo: dall'economia italiana all'economia europea* (1998 – prot. 9813030947_002, Catholic University Research Unit) made possible the completion of my research.

I would like to thank Marina, David, Marco and Matteo Jones, Carlos, Yoya, and Clara Izsak for their hospitality and friendship. I am thankful to Simona Beretta, Enrica Maggioni, Guido Merzoni, Mario Nosvelli, Massimiliano Riggi, and Erika Uberti who have given me personal and moral support. I am also indebted to my parents, Pinuccia and Tarcisio, and to my brother Luca for their constant encouragement.

Last but not least, Nani, my wife, has carried a part of the burden of the work since its beginning and she has had the "better and worse" (and, above all, the worse) of the whole process. Marta Brigid, my daughter, has ranked the University of Warwick Crèche "the best kindergarten in the world". However I am pretty sure that she will be happy to spend a summer at the sea-side without competing with a thesis for her father's time. To the two of them I express my love and gratitude.

Few years passed since the completion of my Ph.D. thesis and the Viva held at the University of Warwick, and few things changed: Anna Lia (my second daughter) was born in March 2000, I was appointed *professore associato* at the Faculty of Political Science, Catholic University of Milan. In the meantime I began to discover that I was referring more and more to my thesis as a personal useful "reference book" for post graduate lecturing and Ph.D. students supervision as well as a background reading and a research agenda for future papers. For this reason, and with the help and encouragement of many people, I decided to write this book. Finally I want to conclude these "acknowledgements – part II" with a special thank to my wife Nani for both her invaluable editorial advice and constant personal support.

M. A. M., Milano, October 2001

Table of contents

Chapter 1 - Introduction **1**
Chapter 2 - What do we know about the clustering of high-tech firms? **5**
2.1 Do firms cluster? 5
2.2 How can clustering be detected? 6
2.2.1 Simple counts, percentages and location maps 7
2.2.2 Relative vs. absolute effects 10
2.2.3 Measures of spatial concentration and inequality 11
2.2.4 The definition of clusters 15
2.3 Where do high-tech firms cluster? 20
2.3.1 Existing empirical evidence 20
2.3.2 New empirical evidence 22
2.3.3 Final remarks and observations 58
2.4 Appendix: data sources 59
Chapter 3 - Theoretical literature review **63**
3.1 Introduction 63
3.2 "Classical" location theory 64
3.2.1 Least cost approach 64
3.2.2 "Demand side" approach 69
3.2.3 "Land utilisation" approach 70
3.3 General location equilibrium models and urban economic theory 71
3.3.1 Non-price interactions models 72
3.3.2 Monopolistic competition models 74
3.4 Game theory 76
3.5 Krugman and the new economic geography 77
3.6 Industrial geography 80
3.7 Technological infrastructure approach 82
3.8 Porter's competitive advantages 83
3.9 Jacobs and the economic history of cities 83
3.10 Diffusion theory 84
3.11 Path dependence, lock-in and informational cascades 86
3.12 Biological and ecological models 88

3.13 Concluding remarks 89
3.14 Moving forward 92
Chapter 4 - Modelling firms' location and cluster development 95
4.1 Introduction 95
4.2 Firm location processes and the development of clusters: a macro-economic (ecologically-derived) approach 96
4.2.1 The effect of previous locations on clustering dynamics 96
4.2.2 Locational benefits and costs and the development of an industrial cluster 97
4.2.3 A logistic model 100
4.2.4 Location process in a two-regions framework 106
4.2.5 Co-operative interactions 110
4.2.6 Multi-cluster interactions: some methodological remarks 115
4.3 Micro-economic foundation of ecological models: the role of expectations 117
4.3.1 Introduction 117
4.3.2 Subjective heterogeneity: the role of expectations 118
4.3.3 Entry dynamics 123
4.3.4 The initial critical mass 125
4.3.5 Firms 126
4.3.6 Exit dynamics 126
4.3.7 Concluding remarks 128
4.4 Cluster development policies in variable macro-economic conditions 129
4.4.1 Introduction 129
4.4.2 Policy interventions for local development 130
4.4.3 The relevance of external macro-economic conditions 131
4.4.4 Variable macro-economic conditions 133
4.4.5 Optimal long run development policies 134
4.4.6 Concluding remarks 136
4.5 Firm location processes and cluster's development: a micro-economic (innovation diffusion derived) approach 137
4.5.1 Epidemic models 137
4.5.2 Rank models 140
4.5.3 Rank, stock and order effects in the location process 143

Chapter 5 - Survey of empirical literature **149**
5.1 Why do firms cluster? 149
5.1.1 Agglomeration economies 151
5.1.2 Localised sources of demand and/or supply 152
5.1.3 Search costs reduction, diffusion, and informational cascades 153
5.1.4 Increase in market power 154
5.2 How do firms cluster? 154
5.2.1 Relevance of innovative infrastructure 156
5.2.2 Relevance of urbanisation/agglomeration economies 157
5.2.3 Relevance of institutional framework and national characteristics 157
5.2.4 Relevance of the availability of skilled labour 158
5.2.5 Relevance of venture capital and financial institutions 158
5.2.6 Relevance of the quality of life 159
5.2.7 Relevance of the proximity of R&D plants to head quarters locations 159
5.2.8 Relevance of site reputation, prestige and business climate 160
5.2.9 Excellence and success do not necessarily bring success 161
5.3 How to study clustering? 161
5.3.1 Advantages and drawbacks of direct surveys 161
5.3.2 Advantages and drawbacks of statistical macro-level studies 162
5.3.3 Advantages and drawbacks of econometric analyses 163
5.3.4 Advantages and drawbacks of simulations 163
5.4 Final remarks 163
Chapter 6 - Empirical analyses of the location of high-tech firms and of cluster development **165**
6.1 Introduction 165
6.2 The role of geographical benefits 166
6.2.1 Estimating the role of geographical benefits in determining location decisions 168
6.3 Clustering of firms or workers? Scale vs. agglomeration economies 172
6.3.1 A rough indicator 173
6.3.2 A model 175
6.3.3 The data 176
6.3.4 The results 177
6.4 The growth of an isolated cluster 179
6.5 Competition and co-operation in the dynamic of clusters development 180
6.5.1 A model 181
6.6. Market structure and industrial specialization in the growth of high-tech clusters 186
6.6.1 A model 188

6.6.2 The data ... 190
6.6.3 The results ... 192
6.7 Why do high-tech firms cluster? Some empirical considerations ... 194
Chapter 7 - Policy implications ... 197
7.1 High-tech firms and local development policies ... 197
7.2 Geographical versus agglomeration benefits ... 199
7.3 Geographical benefits and firms location ... 200
7.4 Agglomeration benefits and firms location ... 202
7.5 Entry versus growth policies ... 205
7.6 The nature of the cluster (and its policy implications) ... 207
7.7 Policy instruments in actions: science parks and technopolis ... 212
7.8 Guidelines for policy makers ... 215
Chapter 8 - Conclusion and research agenda ... 219
References ... 221
Declaration ... 237
Summary ... 239
List of figures ... 241
List of tables ... 243

Chapter 1 - Introduction

> Across a line drawn from New York to Los Angeles, the level of economic activity is hardly uniform. In principle, the regional economist ought to be able to predict the agglomeration of activity at certain points (...).
>
> R.E. Hall (1991), *Booms and Recessions in a Noisy Economy.*

This book aims to study the location process of high-technology firms and to analyse the emergence of spatial clustering in such innovative sectors.

The relevance of the issue at study is witnessed by its current centrality in the theoretical and empirical literature, and in the policy debate. From a theoretical viewpoint, the analysis of spatial problems (of "geography and trade") has recently been re-admitted to the realm of economic theory after long years of exile. From an empirical perspective, the analysis of spatial knowledge spillovers, together with inter-industrial ones is getting much attention from the scholars. From a policy-oriented standpoint, the current world-wide globalisation process, together with continental processes of integration, has gradually but crucially shifted the focus far from the national level toward the two extremes: the regions and the world.

The book is structured into eight chapters.

The second chapter is dedicated to a detailed analysis of the problem. In the first sections of the chapter the empirical relevance of the issue is firstly discussed. Then a series of statistical and economic indicators, identifying and measuring spatial industrial clustering, are illustrated and applied to an original data-base, which has been built for the book (Maggioni 1999), concerning four major industrialised countries (US, UK, France and Italy).

The third chapter is devoted to a particular kind of literature survey. Different streams and also different disciplines have been reviewed in order to highlight the theoretical underpinnings of firm's location decisions and the dynamics of industrial clustering. In particular the first section is dedicated to the contributions put forward by the founding fathers of location theory and to the analysis of three main approaches proposed by the so-called classical school: namely "least cost", "demand side" and "land utilisation" approaches. Subsequently the general equilibrium representation of the location problem has been examined, with specific reference to non-price interactions and monopolistic competition models. Strictly connected to the previous section, a further section is devoted to the literature originating from the contributions of Krugman, which are based on the introduction of increasing returns and imperfect competition into the neo-classical framework of general equilibrium. Another section deals with the "industrial geography" approach, which sees the economic regions as the product of the striving forces of industrial capitalism. This approach is followed by a survey of the "technological infrastructure approach", which stresses the relevance of scientific and technological agglomeration economies in determining the innovative performance of a region and its industrial specialisation. Two sections are respectively devoted to a brief survey of Porter's and Jacobs' contributes. These two authors, who belongs to the neighbouring scientific communities of strategic management theorists and economic historians, have acutely highlighted some interesting features of the process of development of industrial clusters. Another section deals with Arthur's contribution to the theory of industrial location and shows some interesting similarities and differences between this approach and the analysis of informational cascades. The final section of the literature survey looks at a peculiar approach to locational issue: the biological and ecological models, which deal with the locational process in a systemic framework. The chapter is concluded by a table which summarises the advantages and drawbacks of each approach when dealing with the topic of study.

The fourth chapter is dedicated to the theoretical definition of a modelling framework able to explain the stylised fact of firm's location process and high-tech clustering. In particular the section 4.2 concentrates — from a macro— economic perspective — on the development path of industries and regions in order to explain two major questions: why do industrial clusters not grow infinitely and why do some cluster grow and others stay small and, sometimes, disappear. Section 4.3 introduces the role of firm expectations into the ecological modelling framework in an attempt to build some explicit micro-foundation for this class of models. A first appendix (section 4.4) analyses the effects of changes in the macro-economic conditions at the national and international level on the development of the cluster and discusses the desirability of different local industrial policies. A second appendix (section 4.5) focuses — from a micro-economic perspective — on the application of models derived from diffusion theory to the location process in order to answer the following questions: Why is location a lengthy process? Why is development generally S-shaped? Why does it display significant variance across industries, regions and countries?

The fifth chapter contains a brief review of the empirical literature on the geographical and agglomeration factors which explain the existence, the location and the growth of high-tech clusters. A section devoted to a discussion of the main methodological techniques (empirical surveys, statistical studies, econometric analyses and simulations) available to investigate the structure and the dynamics of innovative industrial clustering concludes the chapter.

The sixth chapter contains the empirical analysis of the book and makes extensive use of the original data-set. In particular, after an introduction, the second section is devoted to the identification of the most relevant locational factors in explaining the location decision of US high-tech firms over the last decade. The third section analyses the relative importance of scale versus agglomeration economies in explaining the industrial specialisation of four major countries. In the same section the role of industrial and geographical spillovers is also explicitly examined. Sections 6.4 and 6.5 show the empirical potentialities and drawbacks of ecologically derived models in the analysis of the processes of firms location and of the development of industrial clusters. The sixth section builds a framework of analysis for testing the empirical relevance of three possible interpretations on how technological externalities and knowledge spillovers cause the growth of high-tech clusters. A final section summarises the empirical findings and tries to bring together the different and sometimes contrasting empirical results within an encompassing framework.

The seventh chapter presents the policy implications which derive from the analysis performed in the book. These are especially relevant for two main reasons. The first refers to the growing regional inequalities across regions. The second is the recent trends in EU economic policy and structural intervention programmes, which have shifted their focus from a sectoral to a territorial one. In particular, after an introduction, sections 7.2, 7.3 and 7.4 analyse, from a policy perspective, the trade-off which exists between a development programme focused on geographical as opposed to agglomeration benefits. The fifth section underlines the issue of entry versus growth supporting policies, while section 7.6. shifts back into theory and addresses the fundamental question of the nature of an high-tech cluster. The seventh section applies the above analysis to a policy instrument which in the recent years has been diffusely used, misused and abused: the science park. A final section, with a bit of immodesty, tries to state a series of guidelines for public authorities which already deal and will deal, even more in the future, with the dynamics of industrial innovative clusters.

A final chapter summarises the results of the book, and set the future research agenda.

Chapter 2 - What do we know about the clustering of high-tech firms?

> Regional or "spatial" economics can be summed up in the question: "What is where and why (...)?" Where refers to location in relation to other economic activity; it involves questions of proximity, concentration, dispersion, and similarity or disparity of spatial patterns (...). Until fairly recently, traditional economists ignored the where question altogether, finding plenty of problems to occupy them without giving any spatial dimensions to their analysis. Traditional geographers, though directly concerned with what is where, lacked any real technique of explanation in terms of human behaviour and institutions to supply the why and resorted to mere description and mapping.
>
> E.M. Hoover (1971), *An Introduction to Regional Economics.*

2.1 Do firms cluster?

From a theoretical viewpoint, when observing the spatial distribution of a phenomenon (say the location of firms) in a given territory (say a nation or a region), three main structures can emerge: clustering (i.e. most firms tend to concentrate in a single or in a few locations), avoidance (i.e. all firms tend to be uniformly scattered, in order to maximise inter-firm distances) and independence (i.e. no clear spatial pattern is visible, locations are as if determined by a random process). A specific branch of statistics[1] is devoted to the identification of specific patterns of distributions of events over a plane which can be tested against the null hypothesis of "complete spatial randomness". Such an hypothesis implies (i) that the intensity of events (in our case firms' locations) does not vary over the plane and (ii) that there are no interactions among events[2]. It is easy to see that the location of firms

[1] Namely the statistical analysis of spatial point patterns.

[2] The independence assumption would be violated if the existence of an event at a given point either encouraged or inhibited the occurrence of other events in the neighbourhood of that point.

does violate both hypotheses since (i) there are considerable variations in the spatial distribution of firms and (ii) the previous location of firms at a given site is likely to influence (in various ways, both positively and negatively) the location of other firms in the same and in neighbouring areas.

From an empirical point of view, firms generally do cluster, and they often cluster according to industry. In every country there is plenty of evidence of the phenomenon of local concentration of specific types of firms due to a plurality of different causes (historical events, knowledge spillovers, availability of raw materials etc.).

Moreover there is a conventional wisdom[3] in the economic literature that high-tech firms are even more likely to cluster than other types of firms because of the relevance that agglomeration economies and knowledge spillovers play in these industries. Finally this process, which is often spontaneous, may be fostered by specific industrial and territorial policy interventions, e.g. the creation of science parks, in order to achieve traditional economic policy targets such as the reductions of unemployment, the promotion of economic and social cohesion, the improvement of the international competitiveness of the national economic system.

This chapter aims at showing the empirical relevance of the issue discussed in the book. In order to accomplish this task we firstly analyse a series of graphical tools and statistical indexes which are used in order to detect and measure the existence and the size of clustering phenomena. The existing empirical evidence is briefly reviewed and then new empirical evidence relating to four developed countries is presented.

2.2 How can clustering be detected?

The phenomenon of industrial clustering[4] can be roughly illustrated by a simple geographical re-aggregation[5] of a general industrial data set — such as: the US *County business patterns*, the British *Census of production*, the French *Enquête annuelle d'entreprise regionalisée* and the Italian *Censimento generale dell'industria, commercio, servizi e artigianato* — in order to observe the spatial distribution of X_{ir}, where X is the measured variable (this being the number of

[3] Although recently disputed by Krugman (1991a and 1991b).

[4] Although in the literature clustering and agglomeration are often used as synonymies, in the book the first will be used to define a static phenomenon of a higher than average spatial concentration of a given type of firms, while the second will be used in conjunction with the concepts of economies and diseconomies to define a dynamic phenomenon which can cause the existence of industrial clusters.

[5] However it is not so simple to obtain a geographical description of the productive structure of a Country. For the UK, for example, few data are published (in official ONS publications such as: *Regional Trends, Local Authority District Analysis of UK Businesses,* and *Business Monitor*) and often these data are aggregated at a geographical or a industry level which does not fit the scope of the analysis. For some countries (and in particular for the UK and France) it has thus been essential to acquire unpublished data.

establishments, the employment level or the sales value) in industry i, in the geographical sub-unit r.

The use of the letter r to define the geographical unit under study derives from the concept of region which will be used, throughout this work, to define a generic administrative sub-unit of a Country[6]. In this chapter, however, it will be used either the official national denominations (such as County and Region, for the UK; *Provincia* and *Regione*, for Italy), or an artificial *ad-hoc* classification composed by FLAs and SLAs. First level area (FLA) defines: *Provincia* (I), County (UK), State (US) and *Département* (F); second level area (SLA) defines: *Regione* (I), Region (UK), Census Division (US), *Region* (F). FLAs and SLAs have been created because of the lack of a coherent international hierarchic definition of geographical areas[7]. The classification used in the book is therefore based on a simple principle. FLA is defined as the first coherent level of an analysis of economies and diseconomies of agglomeration and is often used by policy maker as the first level of economic policy. SLA allows one to consider other sources which can explain the structure of industrial clustering by taking into account some macro factors and existing geographical differences within a country in the labour force skills, business climate, physical and informational infrastructures (such as the Italian or British "North-South divide", the US "industrial belts").

2.2.1 Simple counts, percentages and location maps

Simply by looking at spatially re-arranged data sets, one may have a feeling of whether or not a relevant number of firms and workers belonging to certain industries are concentrated in some specific geographical area. One may further compare the local amount of high-tech activities with the national average $\overline{X}_{ir} = \frac{X_{iR}}{N}$, where N is the number of regions in the larger geographical area of reference (usually the nation) which is called R, and $X_{iR} = \sum_{r=1}^{N} X_{ir}$.

One can also calculate the local percentage (of the national industry total) through the following ratio:

$$AX_{ir} = \frac{X_{ir}}{X_{iR}}; \qquad (2.1)$$

in order to measure the relative distribution of an industry in different regions[8].

Alternatively, if one plots the data (on establishments, employment or sales) regarding a specific industry or group of industries on the map of a country, and

[6] While the term cluster will be used to identify the economic entity which is defined by the combination of an industry and a region.

[7] EU has its own threefold geographical classification (NUTS) but this does not correspond to the single country definitions of administrative units (i.e. NUTS2 in Italy are not regions but sets of regions) and, by definition, does not take into account the US.

[8] This percentage is a building block of several concentration indexes which will be illustrated in section 2.2.3.

uses some simple graphical techniques (as using darker shades to indicate the presence of a stronger phenomenon or spot of different dimensions, where the dimension of the spot is proportional to the size of the variable under study), then the emergence of spatial concentration of X_{ir} can be made graphically evident. Figure 2.1 and 2.2 illustrate the use of such techniques[9].

[9] Throughout the book, US maps do not include Alaska and Hawaii in order to obtain a larger picture of "continental" USA.

Figure 2.1 Geographical distribution of integrated circuit manufacturing establishments (1982)

Source: Scott (1988a)

Figure 2.2 Location of US manufacturing employment (1990)

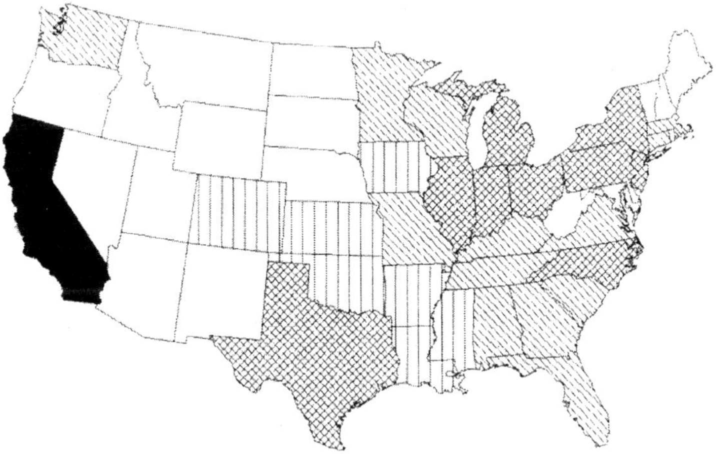

Source: Harrington and Warf (1995)

However the simple geographical descriptions offered by such location maps cannot be taken as indisputable evidence of the existence of industrial clustering, for they can be criticised on a number of grounds.

2.2.2 Relative vs. absolute effects

The biased distribution of the percentages of establishments and employment towards certain locations can be simply due to the fact that some areas are larger and/or more densely populated than others. In order to achieve a more precise description of the clustering phenomenon one should thus utilise some slightly more sophisticated indexes.

The easiest and most popular "trick", to solve the area size-bias problem, is to weight the number of firms and/or the level of regional employment in industry i by the size of the local population P_r. In this way, clustering in larger and densely populated areas is no longer overestimated. One then looks at PX_{ir}, where

$$PX_{ir} = \frac{X_{ir}}{P_r} \qquad (2.2)$$

However, when using this normalisation technique one must take into account that the spatial distribution of a population is not an entirely exogenous variable; on the contrary an industrially developed area is likely to have a density of population above the average because of the "labour market pull" phenomenon. One must also consider that, especially at low geographical levels, as in FLAs, such a technique does not take into account commuting and can therefore result in highly biased measurements.

Alternatively, one can use as weights the local amount of manufacturing employment, M_{Ir}, where I is the total of all manufacturing industries i. This last measure seems preferable because it takes into account the "manufacturing" size of the region and avoids the underestimation of particular locations (such as, for example, the Silicon Valley) where there is either a feeble manufacturing tradition (and a strong primary sector) or an extremely developed tertiary sector, and the only manufacturing activities in site belong to high-tech industries. Formally one then uses:

$$MX_{ir} = \frac{X_{ir}}{M_{Ir}} \qquad (2.3)$$

A more subtle technique relates to the use of location quotient, LQ_{irR}, which allows one to compare the specialisation of an area with respect to a larger area unit (which is used as a reference). More formally, if one wants to measure the relative specialisation of region r in industry i with respect to a larger area R (say the nation) and the total of manufacturing industries I, by using the variable X (i.e. number of firms, size of employment, sales), then the location quotient can be written as follows:

$$LQ_{irR} = \frac{\dfrac{X_{ir}}{X_{Ir}}}{\dfrac{X_{iR}}{X_{IR}}} \tag{2.4}$$

The location quotient[10] measures the ratio between the industrial specialisation of the smaller area in comparison with that of the greater one; when its value is greater than one, then the local economy is more specialised (in that particular sector) than the larger economy; the opposite holds if the value is smaller than one.

However it must be noted that location quotients — as with any other relative measure of specialisation — are not substitutes for absolute indexes, for the two give complementary information. Absolute indexes and their graphical counterparts (the so called location maps) show the localised consistence of a phenomenon; while relative indexes on the one hand may help to eliminate size related biases but, on the other, run the risk of overestimating specialisation in case of negligible absolute relevance. Furthermore the majority of micro phenomena involved in agglomeration dynamics have an intrinsically non linear nature, where threshold effects and critical sizes play a major role. For these reasons the absolute value of a localised variable can give more analytical insights than other more sophisticated relative indexes.

2.2.3 Measures of spatial concentration and inequality

The study of industrial clustering is similar to the analysis of the degree of concentration of an industry. In other words the task of the researcher is similar: he/she must find a uni-dimensional measure, incorporating two relevant aspects of the structure: the number of units and the inequalities in size of such units[11].

Thus the ideal index of spatial concentration SC should be a function of: the number of areas N where the studied phenomenon is present, and the size inequalities Q of the phenomenon across the areas (Waterson 1984), such that:

$$SC = f(N,Q); \qquad f_N < 0, f_Q > 0$$

for it is natural to assume that the smaller the number of areas where a given industry is present and the more unequal the distribution of the industrial location across these areas, the more spatially concentrated is the industry. There is an en-

[10] One may further note that this quotient has different names in different streams of literature. In the international economics literature, where the measured variable is the value of exports, it is called Balassa index or index of revealed comparative advantage; in the applied industrial economics literature, where the measured variable is the number of patents, it is called index of revealed technological advantages or index of comparative technological specialisation (Paci and Usai 1997).

[11] Here size refers to the local amount of the variable under study (number of establishments, employment level, etc.).

compassing "and also enlightening way of thinking of concentration, namely as a weighted sum of (area) shares[12]:

$$SC_i = \sum_{r=1}^{N} s_r g(s_r); \qquad 0 \le g(s_r) \le 1$$

where $g(s_r)$ is a weighting scheme" (Waterson 1984, p. 168).

Among all the possible indexes and measures of spatial concentration (which have been adapted from industrial economics handbooks) my book will focus on the following — concentration ratio, Herfindahl index, coefficient of variation, Gini coefficient (in two different version), and Linda index — which may all be considered as different interpretations of the above mentioned general measure.

The spatial concentration ratio *SCR* is the simplest and — ranking regions by the number of establishments (or level of employment) — is defined as the sum of the shares of industry located in the first *n* regions (where *n* is chosen as a significant threshold for measuring spatial concentration):

$$SCR_{in} = \sum_{r=1}^{n} s_{ir}; \qquad \frac{n}{N} < SCR_{in} \le 1 \tag{2.5}$$

SCR_{in} is a discrete index, for it refers only to a specific point of the cumulative concentration curve (described below). For this reason it is possible that for different values of *n*, a given industry *i* may appear both more and less concentrated than another. Furthermore it appears rather arbitrary (if not useless) to compare SCR_{in} for countries which have different numbers of sub-national areas *N*[13].

The second measure of concentration proposed is the spatial version of the Herfindahl index, which is defined as follows:

$$SH_i = \sum_{r=1}^{N} s_{ir}^2; \qquad \frac{1}{N} \le SH_i \le 1 \tag{2.6}$$

If an industry is equally distributed among all the areas (and therefore we deduce that clustering is not very prevalent), then $SH_i = 1/N$. The more unequal is the geographical distribution of the industry (keeping *N* fixed) the closer the index is to one[14]. The main difference with the standard concentration ratio refers to the fact that in the industrial economics literature the maximum number of firms is an endogenous variable which can take very large values[15]. Here, on the contrary, the number of "populated areas" is upper bounded by the exogenous number of areas (FLAs or SLAs) in which the country is divided. Therefore, for the purposes of

---

[12] The share is calculated as the local amount of the measured variable divided by the national total and is thus equal to Ax_{ir} defined as in (2.1).

[13] It is possible to normalise the value of *SCRin* in order to allow for international comparisons so that the transformed index, *RSCRin*, would vary between 0 and 1 irrespective of the value of *N*. $RSCRin = (SCRin - n/N)/(1 - n/N)$. However the value of such a transformation is at least doubtful since the direct connection with the percentage of the first *n* areas is lost.

[14] The limit being when the whole of an industry is concentrated in one single area, a situation which can be defined, following Arthur's (1990) terminology, as locational monopoly, where $SCR_1 = 1$.

[15] And the average firm dimension is negligible, as in the perfect competition case.

this study, we have calculated a normalised version of the index SNH_i which takes into account the existing differences in N between various countries. Formally:

$$SNH_i = \frac{SH_i - \frac{1}{N}}{\frac{N-1}{N}} \qquad 0 \leq SNH_i \leq 1 \qquad (2.7)$$

This index has the advantage that its value varies between *0* (uniform distribution) and *1* (locational monopoly) irrespective of the number of areas into which the country is divided.

The last concentration measure considered in the book is the Linda index, which is a discrete index based upon the concentration ratio allowing for inequalities between different largely "populated" areas which has been extensively used in EC studies of industrial concentration (Linda 1986). The spatial version of Linda index for the first n regions (ranked as in expression 2.5) can be defined as follows:

$$SL_n = \frac{1}{n(n-1)} \sum_{r=1}^{n-r} \frac{n-r}{r} \frac{SCR_r}{SCR_n - SCR_r} \qquad (2.8)$$

If all regions up to the nth are equally populated, then $SL_n = 1/n$. If the (n + *1*)th region is much less populated than its predecessor, the index SL_{n+1} will be much larger than SL_n. The use of a series of Linda indexes, calculated for different values of n, helps to identify an "industry spatial core"[16] which is defined by the number of regions for which a minimum value of SL_n is reached.

Other measures used in the analysis — which cannot be properly defined as concentration indexes but rather as inequality measures — are the coefficient of variation and the Gini coefficients.

The coefficient of variation is the simplest statistical measure of inequality. It is calculated as the ratio between the standard deviation and the mean of a distribution of a given variable and it is therefore independent of the variable's mean value. Formally, the spatial coefficient of variation can be expressed as follows:

$$SCV_i = \frac{\sigma_{ir}}{\overline{X}_{ir}} \qquad SCV_i \geq 0 \qquad (2.9)$$

If an industry is evenly distributed across areas, then the value of the spatial coefficient of variation is equal to zero. The higher the value of SCV_i the more unequal is the spatial distribution of the industry in a given territory.

The Gini coefficient, which is commonly used to describe the degree of income inequality within a country, can be easily modified in order to measure the degree

[16] The industry spatial core — which can be seen as the spatial counterparts of Linda's original concept of "oligopolistic arena" — can be sensibly identified only if the selected areas (or at least a significant part of them) are spatially contiguous.

of inequality in the spatial distribution of a given country. It is calculated as the ratio of two areas[17] described in figure 2.3.

Figure 2.3 Geometric representation of Gini coefficient

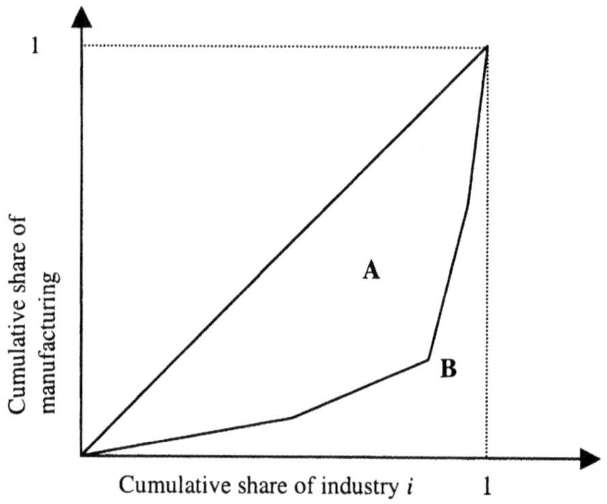

Referring to figure 2.3, the Gini coefficient is equal to $A/(A+B)$; since the area of the triangle (A+B) is, by construction, equal to *1/2*, it follows that $G = \frac{A}{A+B} = 2A = 1 - 2B$.

In this book two versions of the Gini coefficient have been used differing in the definition of "uniform distribution". "Locational Gini coefficient" (*LG*) considers as uniform a spatial distribution of industry *i* such that each area *r* registers exactly *1/N* of the total of the industry; formally:

$$LG_i = \frac{N+1-2\sum_{1}^{N} rs_{ir}}{N}; \qquad 0 \le LG_i \le 1 \qquad (2.10)$$

"Krugman-Gini locational coefficient"[18] (*KG*), considers as uniform a spatial distribution of industry *i* such that each area registers exactly the same shares for industry *i* and for total manufacturing *I*. Formally:

[17] The first being the area enclosed between the cumulative concentration curve (also called locational curve), and the diagonal of the diagram (a 45-degree line), the second being the area of the whole triangle (by construction equal to *1/2*).

$$KG_i = \frac{1}{2}\sum_i \left| \frac{X_{ir}}{X_{Ir}} - \frac{X_{iR}}{X_{IR}} \right| \qquad 0 \le KG_i \le 1 \tag{2.11}$$

By referring to figure 2.3 in order to calculate *KG* we substituted the cumulative number of regions (from *1* to *N*) with the cumulative regional share of total manufacturing. To ensure a better comparability with *LG*, differently from Krugman (1991a), we halved the value of his original index, so that in the book, *KG* varies between *0* and *1*.

A final index refers to the degree of spatial association displayed by different high-tech industries. This index, which is calculated as a correlation coefficient between the regional relative distribution of two industries, allows one to measure the degree of similarities existing between the spatial patterns of a couple of industrial sectors.

$$SA_{ij} = \frac{\sum_r MX_{ir} MX_{jr}}{\sqrt{\left(\sum_r MX_{ir}\right)^2 \left(\sum_r MX_{jr}\right)^2}} \qquad -1 \le SA_{ij} \le 1 \tag{2.12}$$

where MX_{ir} is defined as in (2.3).

When the spatial association coefficient SA_{ij} is close to 1, the industries are highly spatially associated (when one industry records a high value in a region, the other records a high value too); when SA_{ij} is equal to -1, there is spatial avoidance between the industries (when one industry is predominant in a region, the other is absent)[19]. By calculating the average across all high-tech industries within a country, one has a rough indication of the homogeneity or heterogeneity of the locational preferences of high-tech industries.

2.2.4 The definition of clusters

i) Spatial definition

A prime issue relates to the best geographical definition (or the size) of the area in which to detect clusters. In the empirical literature, different sizes have been chosen in order to identify firms clusters.

---

[18] This index has been known for a long time in the regional science literature as "index of localisation" (Florence 1948; Isard 1960). However in the book we call it "Krugman-Gini locational coefficient" since it has been re-discovered by mainstream economics after its use in Krugman (1991a).

[19] This index is thus different from the index of geographic association (Florence 1948) which, being based on the coefficient of localisation, varies between 0 (complete geographic association) and 1 (non geographic association).

If we limit the analysis to standard administrative entities for which data are available, one can study the process of firm location, in the US case, with respect to 4 Regions, 9 Divisions, 51 States[20] or 3126 Counties. In the UK one can base the analysis on 4 Countries, 11 Regions, or 65 Counties[21]. In Italy one can look at the agglomeration issue on the basis of 20 *Regioni* or 95 *Province*[22]. In France one can focus the analysis on 22 *Régions* or 95 *Départements*.

In most cases, the level of the analysis is determined by the availability and the overall manageability[23] of data. In general one must perform the analysis at the territorial level, which corresponds to what geographers would call "homogeneous or uniform region", (i.e. areas characterised by similar industrial structure, demographic patterns and labour market dynamics) and which fits best the topic at hand. However, it is very difficult to find such homogeneous regions in practice (Boudeville 1966). For this reason, later in the chapter some descriptive analyses — aimed at showing the existence and the relevance of clustering of high-tech industries in different countries — have been performed both at the FLA (County in the UK, *Département* in France, *Provincia* in Italy, State in the US) and at the SLA level (i.e. Region in the UK, *Région* in France, *Regione* in Italy, Census Division in the US). The use of SLAs allows one to test whether economies of agglomeration extend their influence over a wider space or, on the contrary, dynamics of locational orphaning prevail[24].

ii) Industrial sector definition

This issue is threefold: it firstly relates to the definition of an industrial cluster *per se*, it also refers to the definition of what high-tech industries are, and finally it concerns the international comparability of different countries' industrial classifications, statistical standards and data collecting procedures.

When one is interested in analysing industrial clustering — which is the product of several interacting forces such as, among others, enjoyment of agglomeration economies, reduction of search costs, exploitation of a localised source of skilled labour[25] — one should carefully consider what is the most appropriate level of analysis (usually expressed in terms of number of digits in the official industrial classification). Wider definitions of industries runs the risk of mixing high-tech

[20] Since we count the District of Columbia as a State.

[21] Since we count Scottish Local Authorities as Counties.

[22] At present in Italy there are 103 *Province*. Since 1991 (date of the last industrial census) 8 new *Province* have been created by splitting old ones.

[23] There is a trade off between the level of geographic disaggregation (and accuracy) and the number of observations which can be reasonably studied.

[24] "When attractiveness varies smoothly over the landscape, then places near a location with a larger number of firms are similar to this location geographically; and they are not sufficiently more attractive to overcome the agglomeration advantage of their dominant neighbour. These places have become dynamically orphaned, and we can say they lie within the agglomeration shadow of their dominant neighbour" (Arthur 1990, p. 240).

[25] For more on this issue, see section 5.2.4.

and not so high-tech activities, narrower ones may prevent the identification and the analysis of technological and productive interdependencies existing between different sub-sectors within the same industry[26].

With regards to the sectoral definition of high-technology activities, in this work we examine the most prominent high-tech sectors according to the OECD (1986) definition which is based on the percentage of sectoral R&D expenditure over total sales, which that should exceed the value of 3%[27]. The high-tech sectors analysed in the book are: Aerospace, Computers and office machinery, Electronic components, Pharmaceuticals and Instruments. The last industry — for all countries, except France — has been divided into 4 homogeneous sub-sectors: Medical instruments, Measurement instruments, Industrial process control instruments, and Optical and photographic equipment.

We built a macro-sector (called total high-tech) as the sum of the previously mentioned industries and — in order to test the claim of Krugman (1991a) that agglomeration is not specific to high-tech industry but, on the contrary, it is displayed more fiercely by "traditional" sectors — we chose two "benchmark sectors" to represent the medium and the low technology industries: Motor vehicles and Textiles. In this way, industries belonging to all types of the Pavitt taxonomy (Pavitt 1984) are included in the analysis[28].

As far as international comparability of data is concerned, we used the European NACE industrial classification at three digit level as standard. In France data at the *Département* level were collected according to an older classification (the NAP, Niveau 100) and therefore a correspondence table between the different classifications had to be constructed (see table 2.1). A similar procedure has been followed to allow the comparability of US data, originally recorded according to SIC (1987 version)[29].

iii) Measured variables definitions

At this point we need to choose the variable to measure industrial concentration. Most of the studies on this issue utilise data on local employment, some of them

[26] This is a debated issue. Swann et al. (1998) use a narrower level of industrial classification, than the one used in the book, arguing that most innovative spillovers develop within (and not between) a NACE 3-digit industry.

[27] Later, OECD (1997) has changed the threshold for high-tech sectors (from 3% to 4%). However this has only marginally changed the list of sectors. Furthermore the list of high-tech industries used in this book has become a conventional wisdom in the empirical literature (see, between others, Acs and Audretsch 1989; Castells and Hall 1994; Feldman 1995; Hall and Markusen 1985; Keeble 1988; Luger and Goldstein 1991; Malecki 1991; Oakey 1981a; Premus 1982).

[28] According to Pavitt (1984) Aerospace, Computers and office machinery, Electronic components, and Pharmaceuticals are science based industries, Instruments belongs to both science based and specialised supplier industries, Motor vehicles is a scale intensive industry, and Textiles is a supplier dominated industry.

[29] For more on this issue see section 2.3.2.

use firms (or local units) count, virtually none uses sales data. The first two variables could be used interchangeably, according to data availability. Very few studies utilise total sales because of the difficulty in obtaining these data at a highly disaggregated territorial level.

However it is easy to understand that, although these variables are highly and positively correlated[30], mapping industrial location through three different variables can result in different pictures.

The number of business units (these being either firms or establishments) is utilised to obtain a description of the solidity and vitality of sectoral industrial structure and entrepreneurship[31]. This variable, however, does not take into account any existing difference in firms' sizes. A locational map based on firm numbers can therefore overestimate emerging local clusters of small (often relatively young) firms at the expense of established locations where the average business unit size is higher[32]. Furthermore some relevant phenomena which generate agglomeration economies (such as producer-user relationships, knowledge spillovers, and local competitive spurs) are related to the number of different business units more than to the local size (in terms of employment or sales) of the industry.

The level of reporting units constitutes another issue: most of the data are collected both at firm and establishment[33] levels. We decided to use the number of establishments for three main reasons: first, production takes place at the establishment level and so we can choose the interesting units in case of market-diversified firms; second, the location decision is often taken at the establishment level (and this is especially true for firms in high-tech industry, where the typical organisational structure is flat and non-hierarchical); third, we wanted to avoid the

---

[30] While it is easy to see why employment and sales are correlated (this is a direct consequence of a reasonable assumption of similar productivity of firms in the same sector), it seems more difficult to explain the correlation existing between the number of establishments and other size-related indicators (employment and sales) which may logically appear to be inversely related (on a particular location one can find either one large establishment or several small ones). However it must be noted that, generally, large establishments are surrounded by lots of small local suppliers and sub-contractors, and this is especially true for high-tech firms which need a stable local network of highly specialised suppliers in order to establish an efficient web of producer-user interactions (von Hippel 1988). Thus the existence of one large establishment in a particular site is likely to be associated with the presence of several other (small) ones in the same or neighbouring locations.

[31] A well established stream of industrial economic literature — devoted to the analysis of firms birth and death processes (see, among others, Keeble and Wever 1986; Storey 1982; Vivarelli 1994 and 1995; Audretsch and Vivarelli 1996; and the special issue of *Regional Studies* 1994) — is based on the analysis of these data.

[32] It must be noted, however, that for a significant number of US high-tech industries (Computer and office machinery, Electronic components, Aerospace, and Computer services) the average establishment size has steadily decreased in the last forty years.

[33] Which is sometimes defined also as plant, local unit or reporting unit.

overestimation of metropolitan areas (and especially capital cities) where most firms locate their headquarters[34].

Industrial employment is the size-related variable most widely used in location studies. Through employment data one can better assess both the absolute and the relative size of specific industries at the local and national level. However, when using this variable, one runs the risk of identifying as industrial clusters, the locations of large isolated firms. Furthermore it must be stressed that, as labour is an input to the production process, every measurement built on it does not take into account the existence of differences in productivity. In this way one can overstate the importance of specific locations where technology is particularly labour-intensive[35].

Total sales (or turnover) registered by firms located in the area is probably the best candidate for an "objective" survey of industrial location. Sales are, by definition, an output measure and, although indirectly, they allow one to measure the "industrial competitiveness" of an area. If 80% of the total sales of a specific sector comes from a particular region, then this region (irrespective of firm numbers and employment) must surely be the place where industrial clustering pays. Unfortunately, the difficulty in obtaining such data (which are usually not recorded by the Census) hinders such analysis.

iv) Time framework

The final issue relates to the time framework of the analysis. In this book we perform different empirical exercises, each exercise requiring different types of data. The descriptive statistic analysis contained in this chapter utilises the most recent data for every country of our sample. In chapter 6, the dynamic analysis requires the longest possible time series, while other econometric exercises requires panel data. For these reasons we end up using, for the static analysis, 1991 data for Italy (1991 being the last census year), 1995 unpublished data for both France[36] and the UK[37], and 1994 for the US. The dynamic and the panel data analyses have involved mostly the use of US County Business Patterns[38] — which gives yearly

---

[34] This has been the reason why we were compelled to use the old NAP 100 classification for France. The data classified according to NACE are recorded in terms of firms (and due to the peculiar geographical corporate structure of French industry, more than 80% of all high-tech firms are located in the *Département* of Paris).

[35] This limitation appears less severe if one considers that, especially within the high-tech sectors, the variance in the production function of different firms is reasonably small.

[36] Data for France were obtained by INSEE within a joint research project — between the Italian and the French Research Councils — carried out by the author with Christian Longhi (Latapses-CNRS).

[37] Data for the UK are published by ONS exclusively at Standard Statistical Region Level. For this analysis we expressly acquired data on "count and employment" at County level.

[38] From 1948 to 1975, County business data are available, only in print, in some US public libraries (thus we accessed the Library of the Congress in Washington D.C.), from 1976

data on number of establishments and employees from as early as 1948 at various industrial and geographical disaggregations — together with other sources of data (described in section 2.4).

2.3 Where do high-tech firms cluster?

2.3.1 Existing empirical evidence

In many countries, the spatial concentration of innovative firms favours already established regions, fostering regional economic growth in these advantaged areas, and increasing spatial development gaps. In Italy the innovative activity[39] of two main industrial regions (Piemonte and Lombardia) accounts for the 60% of the national total (Antonelli 1988; Ciciotti 1992; Maggioni and Miglierina 1995). In the UK, the South-East region dominates in the location of high-technology manufacturing and services (Howells 1984; Begg and Cameron 1988; Oakey 1984). In the US, five states (California, Massachusetts, New York, New Jersey and Pennsylvania) account for more than 50% of total shipments of semiconductors and integrated circuits (Scott 1988a; Storper and Walker 1989). A slightly different list (which includes Florida and Illinois and excludes New Jersey and Pennsylvania) records more than 40% of total US establishments in the medical devices sector (Scott 1993). In France, a large and expanding industrial region (called "la Cité Scientifique") which is formed by a loose aggregation of 87 City Councils belonging to four *Départements* in the outskirts of Paris (Yvelines, Hauts de Seine, Val de Marne and Essonne) accounts for more than one third of the French total high-tech employment. In Japan two major metropolitan areas (Tokyo and, in a minor way, Osaka) constitute the core of a very rigid centre-periphery territorial structure for high-tech industries, both for R&D and production departments (Morita and Hiraoka 1988).

A conventional wisdom, which has gradually emerged in the existing empirical literature on the issue, states that high-tech firms cluster in large sub-urban areas close to and easily connected to metropolitan areas. Both production externalities and consumption externalities are responsible for this choice. From a production perspective, innovative firms (which are usually both small and newly established) look for an urban area close to a university (or a research centre) with a good scientific and technological reputation where a highly skilled labour force is easily available, and connected to the rest of the world by modern infrastructures of transport and communication. From a consumption perspective these areas should also be endowed with pleasant residential areas, with the presence of natural and cultural amenities and without congestion phenomena (such as pollution, traffic,

to 1986 data are recorded on ("old fashioned" 9-track 6250 bpi) magnetic computer tapes, from 1986 onwards data are available on CDs.

[39] Measured by the number of patents, the R&D employment and the R&D expenditure.

noise, criminality) typical of the inner cities of a large metropolis. However it must be considered that — because of the existence of agglomeration economies and locally increasing returns to location — the exact positioning of an industrial cluster within a region may well have been initially determined by chance or by historical accidents to become later established because of the cumulative nature of firms' location processes.

The applied empirical literature on the issue is rich in case histories which try to identify the reasons for success of famous high-tech cluster such as: Santa Clara County (US), and Route 128 (US) for the semiconductor and computer industries; Orange County (US), West Yorkshire (UK), and Mirandola (I)[40] for bio-medical apparatus and medical instruments; Hertfordshire (UK) and the North Carolina "triangle"[41] (US) for the pharmaceutical industry; Toulouse (F) and Seattle (US) for aerospace industry.

Some of the above mentioned experiences may be defined as "spontaneous" clusters since they were generated and developed by a series of autonomous location decisions. Others clusters are either the pure product of the will of a regional planner (*in vitro* location) or, more often, a combination of local authorities planning and individual firms autonomous decisions. There are many different names which have been used, in the literature, to define these "artificial clusters" such as: technopolis, science parks, innovation centres, etc. In these cases, clustering is not the sole result of autonomous entrepreneurial choices, but also the response to a planned structure of locational incentives.

Nowadays the support to science and technology activities is in on policy agenda of any public decision maker (both at local and at national/super-national level). The fostering of innovative industrial clusters is one of the main instruments through which such policy may be effectively implemented. It is thus becoming more and more difficult to empirically distinguish between spontaneous and artificial clusters[42].

Section 2.3.2 is devoted to a detailed presentation of the results obtained by the author using an original database for studying the geographical localisation of high-tech industry in four major industrialised countries (USA, UK, France, and Italy[43]). The analysis is conducted in such a way as to identify the major high-tech clusters and to empirically demonstrate that high-tech firms do cluster. The aim of the entire book, is to go further and to also try to discover how and why firms cluster.

[40] However it must be noted that typical Italian high-tech clusters are somehow dissimilar from the ones existing in other countries. Italian cluster are based on a lower level of technology and they consist in a new variety of the "industrial district model", where medium and advanced technology are adopted and adapted to the needs of mature sectors.

[41] Identified by the cities of Raleigh, Chapel Hill and Durham.

[42] For a theoretical discussion about this issue, see Chapter 7.

[43] Ranked in terms of relative size of total high-tech employment.

2.3.2 New empirical evidence

i) Data description and transformations

This section illustrates the original data set which has been used for the descriptive analysis and highlights the limits to the comparability of the figures.

For all countries data have been recorded on number of establishments and employment at both FLA and SLA levels. Sector identification codes are defined following the EU NACE classification: 244, Pharmaceuticals; 300, Computers and office machinery; 321, Electronic components; 331, Medical and surgical instruments; 332, Measuring and controlling devices, search and navigation instruments; 333, Industrial process control instruments; 334, Optical and photographic equipment; 353, Aerospace; 170, Textiles; 240, Motor vehicles; D, Manufacturing. A macro-sector, denominated total high-tech, has also been built as the sum of all high-tech sectors. Exceptions to this scheme are the following: in France there is only one instrument sector; in the UK most of the figures for employment at county level have been removed by ONS to avoid disclosure[44] and therefore some indexes have been calculated only on establishments data. For the UK we were also unable to get data for the medium and low technology sectors (Textiles and Motor vehicles).

[44] A similar, but minor, problem regards US data. In that case, however, censured employment figures were substituted by size class flags. Differently from Krugman (1991a) we decided not to drop these units from the statistical analysis and we substituted the size class flag with the median of the size class (i.e. the "C" flag, corresponding to a size class of 100-249 employees, has been substituted by a figure of 175). By checking the resulting industry totals, obtained by adding the figures for all States, against the official figure for USA, we discovered a negligible "error" (the average size being equal to 0.64%).

Table 2.1 The original database

Industries	industrial classifications			
	NACE (ITA)	NACE (UK)	NAP (FRA)	SIC (USA)
Pharmaceuticals	244	244	19	283
Computers and office machinery	300	300	27	357
Electronic components	321	321	29	367
Medical and surgical instruments	331	331	-	384
Measurement instruments	332	332	-	381+382-3823
Industrial process control instruments	333	333	-	3823
Optical and photogaphic equipment	334	334	-	385+386
Instruments	330	330	34	380
Aerospace	353	353	33	372 +376
Total high-tech	H-T	H-T	H-T	H-T
Textiles	170	-	44	22
Motor vehicles	340	-	31	371
Total manufacturing	D	D	D	20

ii) Absolute and relative measures of industrial location

United States of America

The data on the USA (which, with a total figure of 23,455 high-tech establishments and 2,328,400 high-tech employees, represent the largest national data set in our sample) confirms the conventional wisdom of a rather polarised structure. The first two positions in the absolute rankings for high-tech sectors record only 11 different State names (California, Florida, Illinois, Massachusetts, Minnesota, New Jersey, New York, North Carolina, Pennsylvania, Texas, Washington). California is the top location for all high-tech sector[45], New York and New Jersey lead only in two relatively more "traditional" high-tech sectors (Pharmaceuticals and Optical and photographic equipment) while Textiles firms and employment are concentrated in North Carolina and the Motor vehicles industry is mainly concentrated in California, in terms of number of establishments, and in Michigan, in terms of employment. In general, the first ranked State records a level of plants and employment which is about ten times the national average[46]. Looking at industrial percentages, California has about 20% of the national total of all high-tech

[45] And also for total manufacturing.

[46] With the exception of the Californian Computers and electronic components (establishments), being around 15 times larger, and the Michigan Motor vehicle (employment), being 20 times larger than the national average.

sectors (against a corresponding value of 10% as far as total manufacturing is concerned), although the largest percentage of employment in one state refers to Optical and photographic equipment in New York (over 37%)[47], followed by Textiles in North Carolina (31%).

The use of relative measures substantially changes the picture. When one uses total manufacturing employment as a weighting criterion, the number of States which get at least once in the three first positions rises to 20; Massachusetts takes the lead followed by New Jersey and New Hampshire. A similar result is obtained by dividing the measured variable by the State population: 17 States are ranked in the first three positions and New Hampshire takes the lead followed by Connecticut and New Jersey. The similarity of the two relative rankings becomes clearer when the analysis is performed at SLA level; in that case, New England becomes the leading division, followed by Middle Atlantic and Mountains — when total manufacturing is used — or by Pacific and Middle Atlantic — when population is used. These data seem to suggest that, because of California, the centre of the US high-tech activity has now shifted towards the West coast; nevertheless the East coast, in which the majority of these sectors have first developed, is still alive and kicking. Another interpretation refers to the fact that, while the West coast and Mountains States are experiencing a relatively recent "wave" of industrial development which influences a large number of industries, the East coast's industrial structure has already experienced a rationalisation process and a shift from the old manufacturing core towards a selected sub-set of high-tech industries and advanced services. A further explanation regards the long established strength and tradition of East coast universities and the effects of this tradition in terms of knowledge spillovers and the creation of a local pool of skilled workers.

Location quotients, calculated on the number of establishments at the FLA level, show the high degree of specialisation in the total high-tech sector of Massachusetts, followed by New Hampshire and Connecticut. These data substantiate the image of an East Coast which, despite the loss of the leadership in absolute terms in favour of the West Coast and the Mountains Region, still shows an industrial structure specialised in the most innovative sectors[48].

The single industry rankings and values of this index are as shown in table 2.2 The specialisation of Kansas in Aerospace industry is second only to the specialisation of the Carolinas in Textiles.

---

[47] Which can be explained by the presence of Kodak.

[48] This result is confirmed when the same analysis is performed at the SLA level. New England is the Census Division which records the highest location quotient (1.57) for total high-tech industry.

Table 2.2 Location quotients for US at FLA level (establishments)

Industries	*first ranked State*	*LQ*	*second ranked State*	*LQ*
244	New Jersey	2.48	Maryland	2.19
300	Massachussets	2.28	New Hampshire	2.26
321	New Hampshire	2.69	Massachussets	2.22
331	Utah	2.09	Colorado	1.86
332	Massachussets	2.50	New Hampshire	2.42
333	Massachussets	2.20	Connecticut	2.01
334	Massachussets	2.02	Utah	1.84
353	Kansas	5.09	Washington	2.97
H-T	Massachussets	2.02	New Hampshire	1.91
170	North Carolina	6.17	South Carolina	5.10
340	Indiana	2.81	Michigan	2.36

Location quotient maps for both establishments and employment (see figures 2.4 and 2.5) show the prominence of the innovative industrial structure of the South West (and Mountains) part of the country, which on the East cost is counterbalanced only by New England States and (at a lower level) by Florida.

United Kingdom

The UK data (which refers to national figures of 13,790 establishments and 467,285 employees[49] in the high-tech sectors) describe, on the one hand, the almost "monopolistic" concentration of high-tech sectors (and indeed of most of the productive activity) into the Greater London County and in the South East Region; on the other, they show the relative success of the development strategies implemented in Scotland, through the establishment of the so called Silicon Glen. The top position in almost every high-tech sectors is usually occupied by Greater London[50], the exception being (when the measured variable is employment) Kent for Pharmaceuticals, Hertfordshire for Computers, Strathclyde for Electronic components and Optical and photographic equipment, West Midlands for Industrial process control instruments and Lancashire for Aerospace industry. The metropolitan areas of London, Manchester and Birmingham[51], being the larger cities in the country, are often in the top positions as far as the number of firms is concerned. This can be seen as an indirect confirmation of the metropolitan incubator hypothesis, which sees metropolitan areas as the ideal seedbed for new high-tech firms given the presence of knowledge spillovers, specialised services and efficient infrastructures. The county chosen as the preferred location for the various

[49] Since a part of the employment figures were removed by ONS to avoid disclosure, we were able to allocate, to their respective counties and regions, only 375,575 employees which correspond to a share of approximately 80% of the "true" total high-tech employment in the UK.

[50] Usually followed by Hampshire and Hertfordshire.

[51] Which may roughly correspond to the West Midland County.

high-tech sectors usually records an amount of firms and employment which is approximately 6/7 times the national average[52]. It is interesting to note that West Midlands and Greater Manchester (two major industrial counties) are the most prominent location for Industrial process control instruments (when measured through employment). This prominence is certainly due to the presence on site of a dense layer of potential and actual users/customers. The Aerospace sector displays a peculiar clustering pattern because of the role played by government decisions; the major locations are often determined by historical events[53]. The clustering of high-tech activities in the UK seems even stronger when measured at SLA level. The South East is always the most preferred region, followed at a large distance by North West and West Midlands.

The use of relative measures (both population and total manufacturing employment) reduces the prominence of the Greater London area, although the leading positions are often taken by other South Eastern Counties; it allows the emergence of Wales (Powys and Clywd for Pharmaceuticals, Clywd and South Glamorgan for Medical instruments[54], West Glamorgan for Industrial process control instruments, Powys for Optical and photographic equipment, Clywd for Aerospace) and Scotland (Borders, Fyfe and Lothian for Electronic components, Strathclyde for Optical and photographic equipment). This is also confirmed at the SLA level, where the South East regions is, in relative terms, overcome by Scotland, Wales, South West and East Anglia.

The relative distribution of establishments (as in figure 2.6) identify the southern regions as the most specialised in high-tech industries (together with Clwyd, Fife and Lothian), while when employment is the measured variable (as in figure 2.7) Scotland show the success of location and re-location policies and, in the South, the specialisation pattern move westward.

Location quotients, calculated on the number of establishment at the FLA level, show that even if the Greater London County loses leadership in the high-tech industry, some other South East Counties (namely West Sussex and Surrey) hold the first two positions in the sector (see table 2.3). The highest specialisation of the industrial structure is shown by the Isle of Wight and Clywd for Aerospace.

---

[52] Less than half the corresponding figure for the US.

[53] The development of the aerospace industry in the Northern part of the country is an historical lock-in from the World War II when an important location criterion for such strategic industry, together with the established tradition in the Metal products manufacturing and Shipbuilding, was the relatively greater distance from Germany.

[54] Signalling the existence of a Welsh cluster of health care related high-tech activities.

Table 2.3 Location quotients for UK at FLA level (establishments)

Industries	*first ranked County*	*LQ*	*second ranked County*	*LQ*
244	Dumfries and Galloway	3.82	Lothian	3.84
300	Berkshire	3.08	Wilthshire	2.32
321	Fife	3.10	Surrey	2.52
331	West Sussex	2.10	Kent - Oxfordshire	1.59
332	Oxfordshire	2.14	Buckinghamshire	2.11
333	Bedfordshire	3.02	Shropshire	2.90
334	Isle of Wight	3.38	East Sussex	2.53
353	Isle of Wight	4.31	Clywd	3.61
H-T	West Sussex	1.92	Surrey	1.84

At the SLA level the prominence of the southern regions in the innovative sectors, as measured by location quotients, is confirmed. South East (1.25) and East Anglia (1.22) are slightly more specialised than the rest of the country.

France

As far as France is concerned, the absolute data (which refer to national figures of 6,561 high-tech establishments and 528,510 high-tech employees) show the prominence of the *Départements* of and around Paris[55] (Haut de Seine, Essonne, Yvelines and Val de Marne) the exceptions being the *Départements* near the Swiss border (Doubs, Haute Savoie, Jura) for the Instruments sector[56]. The two traditional "benchmark" sectors (Textiles and Motor vehicles) are localised respectively in the *Départements* of Nord and Doubs. However the areas around Paris are in the top positions also in these low and medium technology industries. The French industrial structure is even more spatially polarised than the US, since the first ranked *Département* often records values more than twenty times higher than the sectoral national average, and as few as 9 *Départements* are always in the first three positions[57]. At the SLA level, therefore, it is no surprise that Île de France is always on top of list of the high-tech regions while Rhône Alpes is almost always second, with the exception of Midi Pyrénées and Aquitaine (for Aerospace).

The relative measures of localisation (especially when the weighting criterion is the manufacturing employment) smooth away the peak performance of Île de France and allows the emergence of other localised centres of high-tech activity

[55] Which are parts of the already quoted "Cité de la Science".

[56] It must be considered, however, that French data for Instruments, being collected at a higher digit level, take into account the production of watches (an activity which is no longer considered as high-tech). These data thus are heavily influenced by the existence of a trans-national watch-making cluster, of long established traditions, which is situated in bordering areas between France and Switzerland.

[57] Other high-tech *Départements* which excel in high-tech sectors, without reach any position higher than the third one, are Rhone (Pharmaceuticals and Computers) and Haute Garonne (Aerospace).

such as Cantal, Lot et Garonne for Phamaceuticals; Herault and Territoire de Belfort for Computers; Côte d'Armour for Electronic components; Indre and Hautes Pyrénées for Aerospace. At the SLA level, the picture appears to be a little less unbalanced towards Île de France, with different *Régions* showing their relative specialisation (Centre for Pharmaceuticals; Alsace and Franche Comtée for Computers; Rhone Alpes and Centre for Electronic components; Franche Comtée for Instruments). The localisation ranking as far as Aerospace is concerned is unchanged by the use of relative measures.

Figure 2.8 shows, above all, the prominence of Île de France (around Paris) for the high-tech industry as a whole, while figure 2.9 highlights the relative importance of a number of southern *Départements* which form a sort of discontinuous chain from Alpes Maritimes to Pyrénées Atlantique.

Location quotients, calculated on the number of establishment at the FLA level, show that, when high-tech industries are considered as a whole, Essonne is the more specialised *Département*[58] and that other *Départements*, which belong to the "Cité de la Science", are in the first ranks in many sectors.

Table 2.4 Location quotients for France at FLA level (establishments)

Industries	*first ranked Département*	*LQ*	*second ranked Département*	*LQ*
244	Cantal	5.37	Hauts de Seine	3.82
300	Yvelines	2.48	Val de Marne	2.36
321	Essonne	3.12	Hautes Alpes	2.42
330	Doubs	7.76	Jura	7.56
353	Indre	5.90	Corse	5.71
H-T	Essonne	2.46	Doubs	2.34
170	Tarn	4.83	Ardeche	4.56
340	Nievre	2.33	Sarthe	1.96

Surprisingly, the location quotients at the SLA level, allow Corse to appear as the more specialised *Région* in the Computer, Electronic components, Aerospace, and in total high-tech industries.

Italy

Italy (the data refer to a total of 29,627 high-tech establishments and 300,876 high-tech employees) displays a segmented geographical structure of the high-tech industries which are concentrated in the northern part of the peninsula and mainly located around the major northern cities and Rome. Milan is almost always the *Provincia* with the greatest number of establishments and largest employment, the exceptions being: Torino for Computers, Belluno for Optical and photographic equipment[59], Varese and Torino for Aerospace. The polarisation of the high-tech

[58] The second ranked FLA is not very relevant since in Doubs the almost totality of figure regarding the high-tech industry is composed by watch-making establishments.

[59] However, it must be remembered that a large percentage of this sector in Italy is constituted by the manufacturing of spectacles frames which cannot be considered high-tech products.

industries is witnessed by the fact that the top ranked *Provincia* records figures always twenty time (and sometimes even more) larger than the national average and that only 11 *Province* out of 95 are ranked in the first three positions. While Textiles sector shows a peculiar location strategy; far from large cities[60] and mainly concentrated in the so-called "Third Italy", Motor vehicle industry, similarly to high-tech sectors, is located in the two major industrial cities of Northern Italy: Torino and Milano. At SLA level Lombardia is always the top *Regione* (the rare exceptions being Veneto for Optical and photographic equipment, and Campania for Aerospace employment), while the second position is alternatively occupied by Piemonte (Computers, Electronic components, Measurement instruments, Industrial process control instruments, Aerospace), Lazio (Pharmaceuticals and Aerospace), Emilia Romagna (Medical instruments, Industrial process control instruments). Location quotient maps for the high-tech industry as a whole illustrate the relevance of large metropolitan areas. Figure 2.10, where the measured variable are establishments, describes the existence of other areas of relative specialisation in the North East and even in the South; while figure 2.11, where the measured variable is employment, show the predominance of Roma.

The relative measures of industrial location (the weights being local population and local manufacturing employment) highlight the existence of relative specialisation in minor *Province* such as: Latina (Pharmaceuticals), Trieste and Aosta (Computers) Macerata and Rieti (Electronic components), Forlì, Modena, Trieste and Imperia (Medical and surgical instruments), Trieste (Measurement Instruments), Modena, Piacenza and Genova (Industrial process control instruments), Aosta and Brindisi (Aerospace). A similar picture, with minor changes, emerges from the location quotients at the FLA level. Milano and Torino are rarely present in the first rankings, while the north-east of the country get the leadership[61].

---

[60] The figures which in these tables are attributed to Firenze are mainly located around Prato which, since 1995, has become a separated *Provincia*.

[61] This result can be seen as a signal of an ongoing transformation of the industrial structure of the so called "Third Italy" which is progressively substituting its traditional industries with more advanced ones but keeping constant the organisational model of the "industrial district" (i.e. a group of SMEs spatially concentrated and connected by strong productive, technological and commercial relationships) (Bramanti and Maggioni 1997).

Table 2.5 Location quotients for Italy at FLA level (establishments)

Industries	*first ranked Provincia*	*LQ*	*second ranked Provincia*	*LQ*
244	Latina	4.36	Roma	3.86
300	Trieste	5.04	Torino	3.72
321	Ancona	3.83	Macerata	3.77
331	Trieste	2.23	Pescara	1.98
332	Trieste	3.34	Milano	2.97
333	Piacenza	4.08	Torino	2.71
334	Belluno	52.4	Cagliari	2.34
353	Aosta	16.14	Gorizia	10.01
H-T	Belluno	5.98	Roma	1.7
170	Pistoia	4.9	Firenze	4.11
340	Torino	5.67	Asti	3.91

At the SLA level, Lazio arises as the *Regione* with the most specialised industrial structure in high-tech sectors such as Pharmaceuticals, Computers, Medical instruments and total high-tech, while Lombardia emerges in two sectors connected to the long established manufacturing tradition (Measurement instruments and Industrial process control instruments). Toscana and Piemonte confirm that clustering is a common and old phenomena in economic history which regards all industries. Toscana, since the Middle Age has been the centre of the Italian textiles industry ($LQ = 2.75$), and Piemonte, despite the delocalisation strategies pursued by Fiat in the last twenty-five years, is still the preferred locations for Italian car (and car parts and accessories) manufacturing.

In the following pages location quotients maps, for both establishments and employment, are displayed for each country in the sample.

Figure 2.4 High-tech location quotients for US at FLA level (establishments)

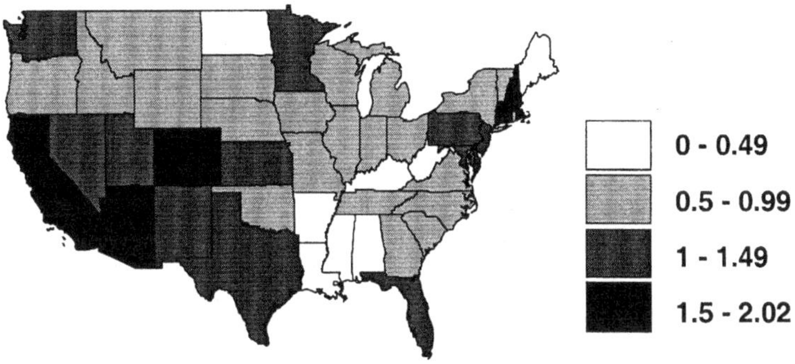

Figure 2.5 High-tech location quotients for US at FLA level (employment)

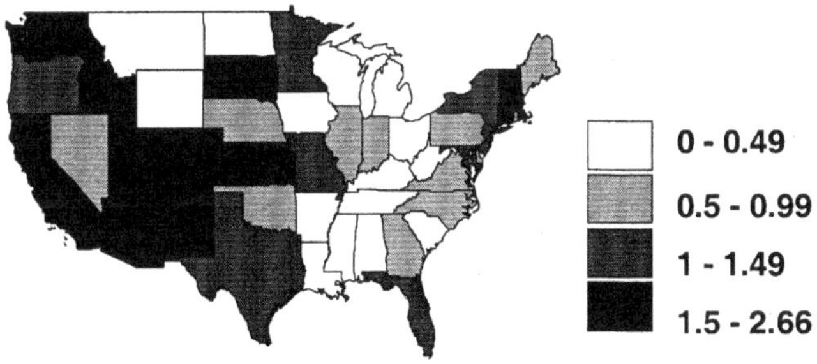

Figure 2.6 High-tech location quotients for UK at FLA level (establishments)

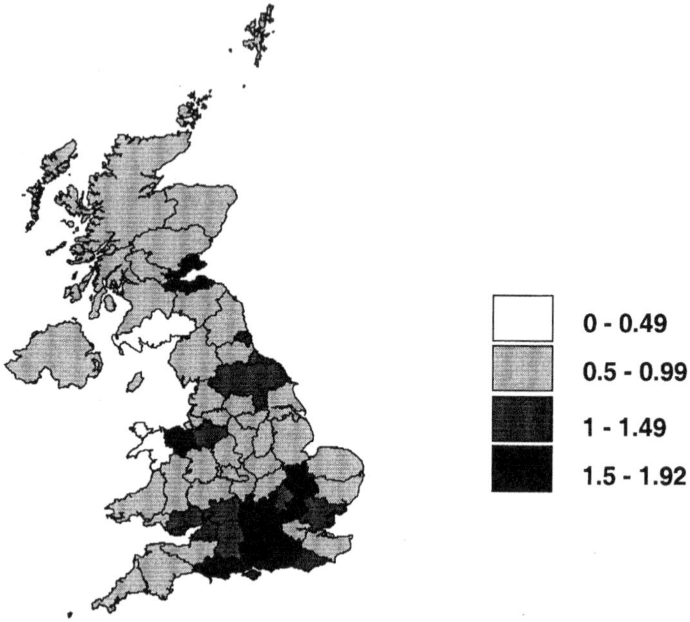

Figure 2.7 High-tech location quotients for UK at FLA level (employment)

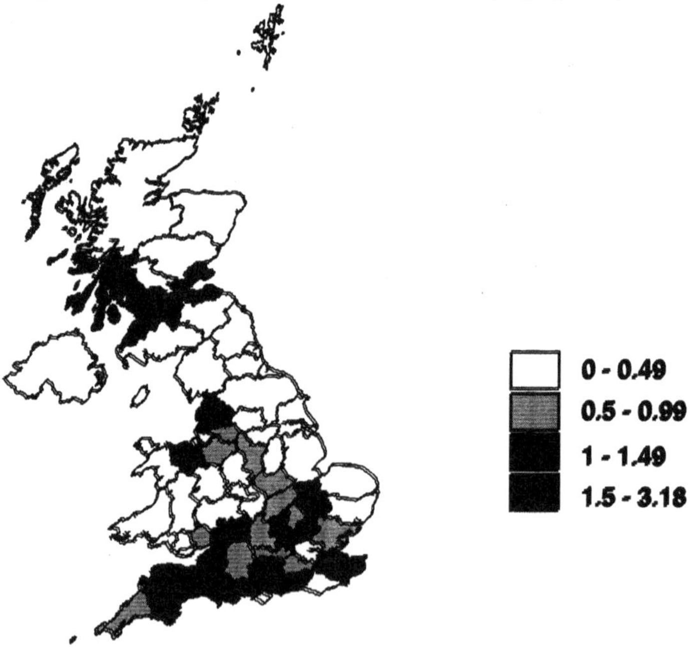

Figure 2.8 High-tech location quotients for France at FLA level (establishments)

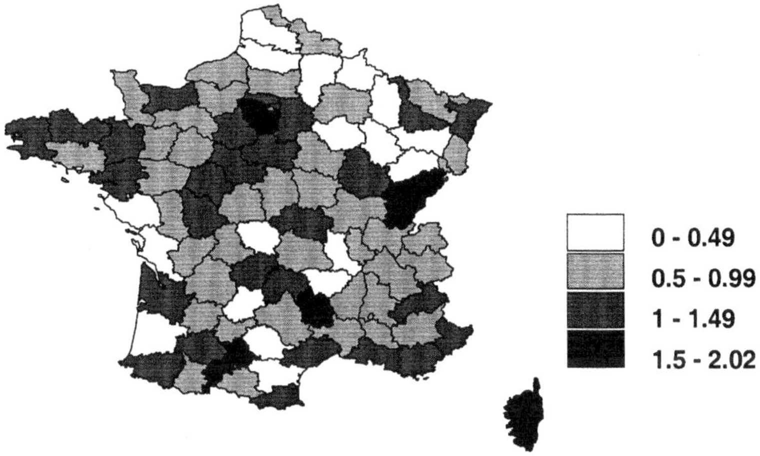

Figure 2.9 High-tech location quotients for France at FLA level (employment)

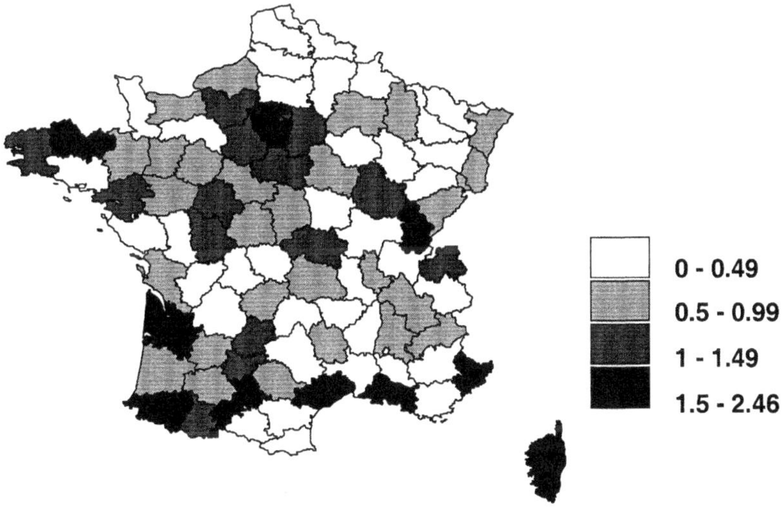

Figure 2.10 High-tech location quotients for Italy at FLA level (establishments)

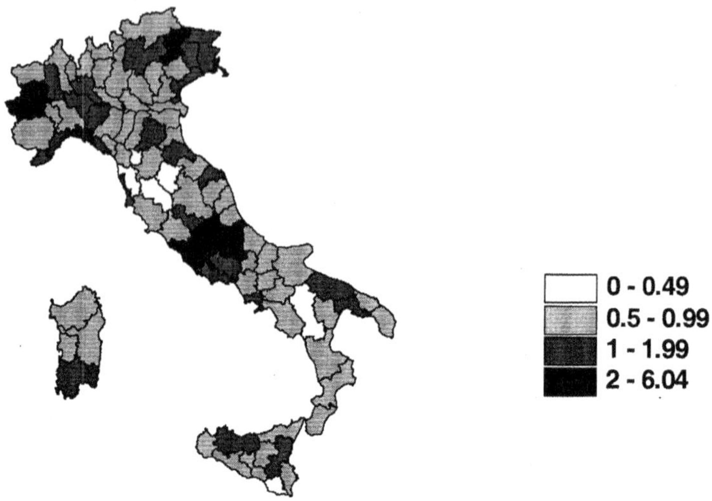

Figure 2.11 High-tech location quotients for Italy at FLA level (employment)

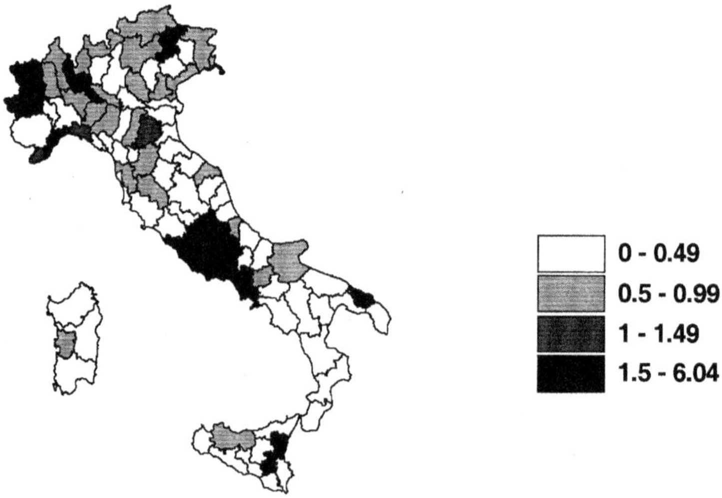

iii) Spatial concentration and inequality indexes

By looking at different spatial concentration and concentration indexes for different industrial sectors, it is possible to extract some hints on the structure and causes of the clustering process.

United States of America

Table 2.6 shows the spatial concentration ratio and inequality indexes for the selected subset of industries studied in the US. It is easy to see that, while according to SCR_4, Textiles are more concentrated, the normalised Herfindhal index shows that Computers and office machinery, and Electronic components are, by far, the most spatially concentrated industries in the US at FLA level.

Table 2.6 US spatial concentration and inequality indexes at FLA level (establishments)[62]

Sectors	SCR_4	Sectors	SNH	Sectors	SCV	Sectors	LG	Sectors	KG
170	0.48	300	0.09	300	2.09	170	0.73	170	0.42
300	0.45	321	0.08	321	1.96	300	0.69	353	0.28
321	0.44	170	0.06	170	1.79	321	0.68	300	0.26
353	0.42	353	0.06	353	1.70	353	0.68	321	0.24
332	0.39	tot h-t	0.06	tot h-t	1.66	333	0.67	332	0.19
tot h-t	0.39	332	0.05	332	1.61	332	0.66	333	0.18
333	0.38	333	0.05	333	1.53	tot h-t	0.65	tot h-t	0.16
334	0.38	331	0.04	331	1.43	334	0.64	334	0.14
244	0.37	334	0.04	334	1.41	331	0.62	244	0.13
340	0.36	244	0.04	244	1.38	244	0.62	340	0.12
331	0.35	340	0.03	340	1.28	340	0.61	331	0.11
D	0.30	D	0.03	D	1.14	D	0.56	D	0.00

The results suggest that Textiles is slightly more concentrated in four States than these high-tech industries.

The same picture emerges from the spatial inequality indexes with Gini coefficients (both versions) which stresses the asymmetric distribution of Textiles while the coefficient of variation underlines the concentration of Computers and Electronic components. What is also evident, is the fact that Motor vehicles establishments are, by far, less concentrated than the almost all high-tech ones[63]. The two

[62] In tables from 2.6 to 2.17, high-tech sectors are made graphically evident through a shaded pattern.

[63] While in Krugman (1991a), who bases the analysis on employment data, "automotive industry is slightly above the median", being ranked as 49th out of 106 manufacturing industries, and is much more concentrated than several high-tech industries such as: Com-

electronic device industries are usually followed by Aerospace, Industrial process control instruments, and Optical and photographic equipment. The health-care related industries (Medical instruments and Pharmaceutical) are ranked almost always in the last positions, showing a more diffused spatial structure.

It is worth looking at table 2.7 where the same exercise is conducted by using employment figures. Table 2.7 is similar to the previous one; however the first ranks are here occupied by those industries in which internal economies of scale play a major role and the average size of establishment is larger.

Table 2.7 US spatial concentration and inequality indexes at FLA level (employment)

Sectors	*SCR4*	Sectors	*SNH*	Sectors	*SCV*	Sectors	*LG*	Sectors	*KG*
170	0.66	334	0.16	334	2.80	170	0.83	170	0.67
334	0.61	170	0.13	170	2.59	334	0.80	334	0.56
353	0.53	353	0.08	353	2.05	353	0.78	353	0.53
340	0.52	300	0.08	300	2.03	340	0.73	300	0.45
300	0.47	340	0.08	340	2.02	333	0.73	340	0.41
321	0.46	321	0.07	321	1.84	244	0.73	333	0.38
244	0.45	332	0.06	332	1.65	300	0.72	244	0.35
333	0.44	244	0.05	244	1.65	321	0.69	321	0.34
332	0.42	333	0.05	333	1.65	332	0.68	332	0.31
tot h-t	0.40	tot h-t	0.05	tot h-t	1.62	331	0.66	tot h-t	0.26
331	0.35	331	0.04	331	1.47	tot h-t	0.65	331	0.23
D	0.26	D	0.02	D	1.03	D	0.57	D	0.00

The concentration and the spatial inequality indexes thus underline the spatial concentration of Textiles, Optical and photographic equipment and Aerospace (although Computers and Electronic components are always in the upper half of the list). Car manufacturing is in the top half of the distribution, confirming the size prominence of the establishments located in Detroit but is generally less spatially concentrated than Computers and office machinery.

When spatial concentration and inequality are looked at SLA level, the general picture stays almost unchanged[64], the main difference being that Textiles and Motor vehicles rise to the top positions[65]. This fact can be explained by thinking that, since it takes time for agglomeration economies and technological spillovers to diffuse over space, the attractiveness of "older" industrial sectors has been able to cover longer distance so to overcome state boundaries and establish Division-wide clusters. However it must be noted that at SLA level some high-tech indus-

puter and office machinery (55th), Pharmaceuticals (70th), Medical and surgical instruments (94th). For a direct comparison, see table 2.7.

[64] Therefore we do not report tables for spatial concentration and inequality indexes calculated at the SLA level.

[65] Their *SCR2* (calculated on the number of establishments) are respectively 0.62 and 0.46.

tries (i.e. Computers and office machinery and Electronic components) do show high values of spatial concentration indexes too[66].

The use of Linda indexes allows the identification of a series of industry spatial cores. For the Pharmaceutical industry (figure 2.12) this core lies in a North-eastern "belt" (which ranges from Missouri to Massachusetts, interrupted in the middle by Ohio) and in three isolated States (California, Texas[67] and North Carolina). The Industrial process control instruments industry (when the measured variables is the employment level) shows a twofold spatial structure (figure 2.13) with a North-eastern belt (from Minnesota to Massachusetts) and a South-western belt (from California to Texas[68]). The Optical and photographic equipment industry (figure 2.14) is concentrated in a wedge-shaped core (with an East-West branch ranging from Minnesota to Massachusetts and a North-south branch which stretches along the entire East coast from Florida to Massachusetts[69]) and three scattered western States (California, Colorado, Texas). The Aerospace industry presents a very dispersed core (figure 2.15) with four clusters: one in the North-east (Michigan, Ohio, New York and Connecticut), one in the South-east (Georgia and Florida), one in the South-centre (Kansas, Oklahoma and Texas), one on the Pacific coast (Washington, California, Arizona). Industry spatial cores arise also in traditional sectors; figure 2.16 shows the endurance of the textiles-based "cotton belt" stretching along the East coast, while figure 2.17 illustrates the resilience (and perhaps revitalisation) of the rust belt", located near the Great Lakes, where car manufacturing and industrial process control instruments industry prevail.

United Kingdom

In the UK, as shown by every index in table 2.8, Pharmaceuticals is the most concentrated industry followed either by Optical and photographic equipment or Computers according to the index used.

---

[66] The spatial concentration ratios (*SCR2*) for Computers and Electronic components are, respectively, 0.45 and 0.44; the spatial coefficients of variation for these sectors are 0.81 and 0.72.

[67] These two States are part of any identified "industry spatial core".

[68] With the exception of New Mexico.

[69] With a gap in correspondence of South Carolina.

Table 2.8 UK spatial concentration and inequality indexes at FLA level (establishments)

Sectors	SCR_4	Sectors	*SNH*	Sectors	*SCV*	Sectors	*LG*	Sectors	*KG*
244	0.31	244	0.04	244	1.60	244	0.54	244	0.17
334	0.30	334	0.03	334	1.37	300	0.53	300	0.12
D	0.29	300	0.03	300	1.28	353	0.52	353	0.11
300	0.28	D	0.02	D	1.22	333	0.51	334	0.09
321	0.27	331	0.02	331	1.17	334	0.50	321	0.08
331	0.26	321	0.02	321	1.16	331	0.48	333	0.07
tot h-t	0.25	tot h-t	0.02	tot h-t	1.09	321	0.48	332	0.04
332	0.23	353	0.02	353	1.00	D	0.47	tot h-t	0.02
353	0.23	332	0.02	332	1.00	tot h-t	0.46	331	0.02
333	0.22	333	0.01	333	0.97	332	0.45	D	0.00

It is interesting to note that, despite the fact that the UK shows a rather polarised spatial structure of production concentrated in the South East region, high-tech sectors are below the figure of all manufacturing sectors (*D*). The locational Gini coefficient[70] confirms this result (although the difference between total manufacturing and high-tech industries becomes almost negligible) and underlines the inequality of the spatial distribution of the Aerospace industry. The lack of employment data for many Counties hinders the calculation of most indexes, with the exception of SCR_4[71]. The analysis of this index strengthens the conclusion that in the UK total manufacturing is almost always more spatially concentrated than high-tech sectors, but also underlines the geographical concentration of the Computer industry which takes first position. It is therefore worth looking at the results at SLA level as shown in table 2.9.

Table 2.9 UK spatial concentration and inequality indexes at SLA level (establishments)

Sectors	SCR_2	Sectors	*SNH*	Sectors	*SCV*	Sectors	*LG*	Sectors	*KG*
244	0.54	300	0.24	300	1.35	353	0.50	244	0.16
300	0.54	321	0.24	321	1.32	300	0.49	353	0.12
321	0.54	244	0.23	244	1.31	333	0.48	321	0.12
331	0.51	334	0.23	334	1.28	334	0.48	300	0.11
332	0.53	332	0.22	332	1.26	244	0.47	334	0.10
333	0.51	tot h-t	0.22	tot h-t	1.23	321	0.47	332	0.09
334	0.53	353	0.20	353	1.14	332	0.47	tot h-t	0.09
353	0.51	331	0.20	331	1.13	tot h-t	0.46	333	0.08
D	0.46	333	0.19	333	1.12	331	0.44	331	0.06
tot h-t	0.52	D	0.17	D	0.96	D	0.41	D	0.00

[70] *KG* is, by definition, equal to zero for total manufacturing.

[71] For this reason, for the UK, tables are presented only with respect to the number of establishments.

At the regional level the situation appears a little blurred, with each indicator showing different rankings for the four main high-tech sectors (244, 300, 321, and 353). According to the concentration ratio, Electronic components take the lead, but both the Herfindhal index and the Coefficient of variation signal Computers as the most concentrated sectors, while the locational Gini coefficient identifies Aerospace and Krugman's version signals Pharmaceuticals. However, it is more interesting that, at this level of analysis, high-tech sectors are more spatially concentrated than total manufacturing. This can be explained by the fact that high-tech industries in the UK are spread in different counties which all belong to the South East and the East Anglia regions, while total manufacturing is more evenly diffused throughout the country.

The Linda system of indexes for the UK most of the times identifies (either a fraction or the total of) the counties belonging to the South East Region plus few other scattered counties (mainly in Scotland or around the cities of Manchester and Birmingham). Some peculiarities are shown by Pharmaceuticals, where there are two cores (figure 2.18) one in the South East of England and the other in the West of Scotland. Computers and office machinery (figure 2.19) shows two distinct core one in the South (from Essex to Wiltshire) one in the west part of Scotland and few scattered Counties in the middle (around Manchester and Birmingham) Electronic components (figure 2.20) displays the presence of a Southern "belt", ranging from Wiltshire to Essex[72] and a Scottish "belt" (composed of Strathclyde, Lothian and Fife). Industrial process control instruments (figure 2.21) are mainly concentrated into two distinct cores, one in the South and one in the central part of England. The first exhibits an inverted Y shape (centred on Greater London with the three branches being respectively in Bedfordshire, Hampshire and Kent). The second core lies in a North-South corridor along the M6 motorway from the West Midlands to Greater Manchester. The spatial distribution of Optical and photographic equipment industry (figure 2.22) confirms the prominence of the South East and highlights the existence of a central "belt" from Hereford and Worcester to Leicester, and a Northern core (Lancashire, Greater Manchester, West Yorkshire). In the Aerospace industry (figure 2.23), the Linda's system of index is almost unable to identify an industry spatial core, since the selected Counties are scattered all around the Country, with the exception of a Northern-central cluster ranging from Lancashire to Derbyshire.

France

In France Textiles is one of the most spatially concentrated industry as can be seen in table 2.10 Within the high-tech industries, Pharmaceuticals and Aerospace record the highest values of concentration indexes, which display almost identical rankings, followed by Instruments and Computers and office machinery.

---

[72] And a number of other scattered Counties such as: Devon, Gwent, Northamptonshire, Greater Manchester, Lancashire, Tyne and Wear, and Northern Ireland.

Table 2.10 France spatial concentration and inequality indexes at FLA level (establishments)

Sectors	*SCR4*	Sectors	*SNH*	Sectors	*SCV*	Sectors	*LG*	Sectors	*KG*
170	0.43	170	0.05	170	2.17	353	0.72	330	0.48
244	0.35	244	0.04	244	1.88	170	0.71	353	0.42
353	0.33	353	0.03	353	1.80	244	0.68	170	0.34
330	0.26	330	0.02	330	1.44	300	0.65	321	0.25
300	0.25	300	0.02	300	1.44	330	0.63	244	0.24
tot h-t	0.24	tot h-t	0.02	tot h-t	1.30	tot h-t	0.57	300	0.16
321	0.23	321	0.02	321	1.27	321	0.55	340	0.16
D	0.22	D	0.01	D	1.12	340	0.47	tot h-t	0.04
340	0.18	340	0.01	340	0.99	D	0.46	D	0.00

As far as inequality indexes are concerned, while spatial coefficients of variation display the same rankings as the concentration indexes, the locational Gini coefficient recognises Aerospace, while Krugman's version identifies Instruments as the most unequally distributed industries. A similar picture is obtained by using employment data, as in table 2.11, the main difference being that Textiles looses its prominence.

Table 2.11 France spatial concentration and inequality indexes at FLA level (employment)

Sectors	*SCR4*	Sectors	*SNH*	Sectors	*SCV*	Sectors	*LG*	Sectors	*KG*
244	0.45	244	0.07	244	2.51	353	0.82	330	0.68
353	0.42	170	0.06	170	2.32	300	0.81	244	0.56
170	0.40	353	0.05	353	2.27	244	0.77	170	0.55
300	0.37	300	0.05	300	2.24	170	0.73	353	0.54
321	0.34	321	0.03	321	1.78	340	0.67	321	0.42
tot h-t	0.31	tot h-t	0.03	tot h-t	1.67	330	0.66	300	0.39
340	0.29	340	0.03	340	1.60	321	0.66	tot h-t	0.38
330	0.28	330	0.02	330	1.52	tot h-t	0.64	340	0.34
D	0.18	D	0.01	D	0.92	D	0.45	D	0.00

At SLA level (displayed in tables 2.12 and 2.13) the most interesting results refer to the clustered structure of the Aerospace industry, which in fact has got more than half of its firms and employment in two single *Régions*[73].

[73] When the measured variable is the number of establishments, the two preferred locations are Île de France and Midi-Pyrénée; when the measured variable is employment, the two preferred locations are Île de France and Aquitaine.

Table 2.12 France spatial concentration and inequality indexes at SLA level (establishments)

Sectors	*SCR2*	Sectors	*SNH*	Sectors	*SCV*	Sectors	*LG*	Sectors	*KG*
244	0.51	353	0.16	353	3.51	353	0.67	330	0.33
353	0.51	244	0.16	244	2.53	170	0.60	353	0.28
321	0.47	321	0.12	300	2.52	244	0.60	170	0.22
total h-t	0.46	total h-t	0.11	total h-t	2.41	300	0.55	300	0.13
300	0.45	300	0.11	330	2.41	330	0.54	244	0.13
170	0.45	170	0.10	170	2.40	321	0.53	340	0.12
330	0.43	330	0.09	321	2.39	total h-t	0.52	321	0.11
D	0.39	D	0.06	D	2.35	D	0.45	total h-t	0.03
340	0.35	340	0.05	340	2.33	340	0.40	D	0.00

It is also interesting to note that at SLA level the high-tech industries as a whole are almost always more spatially clustered than the two benchmark sectors. This phenomenon is more evident in table 2.13, which refers to employment data, and it seems to suggest that in France the optimal scale for the diffusion of productive and technological externalities, which are involved in the dynamics of high-tech clusters, is larger that the *Départements*[74].

Table 2.13 France spatial concentration and inequality indexes at SLA level (employment)

Sectors	*SCR2*	Sectors	*SNH*	Sectors	*SCV*	Sectors	*LG*	Sectors	*KG*
244	0.58	300	0.23	300	2.56	353	0.74	330	0.52
300	0.58	244	0.21	244	2.55	300	0.71	244	0.41
353	0.54	353	0.18	353	2.51	244	0.69	170	0.39
321	0.50	321	0.17	321	2.48	321	0.60	353	0.35
tot h-t	0.49	tot h-t	0.15	tot h-t	2.46	tot h-t	0.59	300	0.29
170	0.46	170	0.09	170	2.42	170	0.58	340	0.28
330	0.38	330	0.07	330	2.37	330	0.57	321	0.20
D	0.33	340	0.05	340	2.34	340	0.49	tot h-t	0.12
340	0.32	D	0.04	D	2.33	D	0.40	D	0.00

The use of the Linda class of indexes in France for several high-tech industries mainly identifies *Départements* which belong to the already mentioned "Cité de la Science" or more generally to the *Région* Parisienne. Another constant presence in these clusters is Rhône or the central part of the Rhône-Alpes region. An exception to this patterns is the Computer and office machinery industry (figure 2.24), whose spatial pattern (when the measured variable is the number of establishments) seems to show the effects of a decentralisation policy, which has added to the *Région* Parisienne a number of other scattered *Départements* — in a wheel-

[74] A complementary explanation refers to the fact that, over the recent years, *Régions* have been in France the most active local authorities in promoting innovation-supporting economic policies. The consequence of such policies may well be the establishment of intra-*Région*, inter-*Département* high-tech cluster.

shape — almost everywhere in the peripheral areas of France (in anti clock-wise order, from Nord to Seine Maritime, from Bretagne to Gironde, from Garonne to Bouches du Rhône, from Rhône-Alpes to Bas-Rhin and Moselle). Aerospace is also scattered and it is possible to identify only a very loose and non contiguous core in the South West. As far as traditional sectors are concerned, the Motor vehicles industry (figure 2.26) is mainly clustered along the Channel coast (from Ille et Villaine to Nord) and along the French-Swiss and German borders (from Moselle to Doubs)[75].

Italy

The analysis of spatial concentration and inequality in Italy, when the measured variable is the number of establishment (see table 2.14), shows a blurred structure where Pharmaceuticals and Aerospace are, according to different indexes, the most concentrated industries, followed by Optical and photographic equipment, and Computers and office machinery.

Table 2.14 Italy spatial concentration and inequality indexes at FLA level (establishments)

Sectors	*SCR4*	Sectors	*SNH*	Sectors	*SCV*	Sectors	*LG*	Sectors	*KG*
244	0.53	244	0.12	244	3.35	353	0.83	353	0.65
300	0.51	334	0.08	334	2.74	244	0.79	244	0.51
353	0.49	300	0.08	300	2.70	300	0.77	300	0.49
333	0.43	332	0.07	332	2.61	333	0.77	333	0.46
321	0.42	353	0.07	353	2.54	321	0.73	321	0.42
332	0.42	333	0.07	333	2.51	170	0.69	170	0.39
334	0.41	321	0.06	321	2.37	332	0.69	334	0.38
340	0.39	340	0.06	340	2.30	340	0.68	340	0.37
170	0.35	170	0.05	170	2.09	334	0.64	332	0.35
tot h-t	0.27	tot h-t	0.02	tot h-t	1.49	tot h-t	0.54	tot h-t	0.13
331	0.25	331	0.02	331	1.33	331	0.51	331	0.08
D	0.20	D	0.01	D	1.09	D	0.47	D	0.00

The inequality indexes identify either Pharmaceuticals or Aerospace as the most clustered industry. In general, the two traditional sectors are less clustered than the individual high-tech sectors but they are more clustered than the high-tech industry when considered as a whole. This apparent puzzle can be explained in term of a scattered clustering structure of each high-tech sector in specific areas[76] so that, when these industries are considered all together as a macro-sector, the value of the concentration and inequality indexes tend to be small.

A similar picture emerges from the employment data (see table 2.15). Here it can be noted that while the Motor vehicles industry rises to the first positions, the Textiles industry (which in Italy is traditionally based on an enormous number of

[75] Other selected *Départements* are those around Paris and Rhône.

[76] This is another indirect evidence of the transfer of the "industrial districts" organisational model from the traditional to the high-tech industries. Each *Provincia*, or even each group of villages, becomes the "leader" of a given production and acquires a monopolistic power on firms locations which extends way beyond the regional borders (locational shadowing).

small firms) falls down to the last rank, being only slightly more clustered than Medical and surgical instruments.

Table 2.15 Italy spatial concentration and inequality indexes at FLA level (employment)

Sectors	*SCR4*	Sectors	*SNH*	Sectors	*SCV*	Sectors	*LG*	Sectors	*KG*
300	0.78	300	0.23	300	4.64	353	0.93	353	0.75
353	0.74	340	0.19	340	4.30	300	0.92	300	0.70
244	0.64	244	0.19	244	4.23	244	0.87	244	0.62
340	0.63	353	0.15	353	3.79	340	0.84	334	0.61
334	0.56	334	0.15	334	3.76	334	0.81	340	0.60
332	0.52	332	0.12	332	3.32	332	0.79	333	0.55
321	0.47	321	0.07	333	3.21	321	0.77	321	0.44
tot h-t	0.46	tot h-t	0.07	321	2.65	tot h-t	0.73	332	0.43
170	0.34	170	0.04	tot h-t	2.64	170	0.73	170	0.37
331	0.32	331	0.03	170	1.89	331	0.60	tot h-t	0.35
D	0.24	D	0.02	331	1.73	D	0.54	331	0.15
333	0.20	333	0.01	D	1.38	333	0.47	D	0.00

At SLA level, the most surprising result, as can be seen in table 2.16, concerns the concentration of establishments belonging to the Industrial process control instruments sector. The first two *Regioni* (Lombardia and Emilia Romagna) record almost 60% of the industry total. This prominence contrasts with the low ranking occupied by the same sector according to the employment figures (table 2.17). This fact can be explained in term of a small average size of the establishments which is a consequence (but also a cause) of the prevailing organisational structure of the industry: the industrial district.

Table 2.16 Italy spatial concentration and inequality indexes at SLA level (establishments)

Sectors	*SCR2*	Sectors	*SNH*	Sectors	*SCV*	Sectors	*LG*	Sectors	*KG*
244	0.58	333	0.19	333	1.94	333	0.75	333	0.38
333	0.58	244	0.18	244	1.89	321	0.71	244	0.36
170	0.51	321	0.15	321	1.75	244	0.71	321	0.33
340	0.51	332	0.15	332	1.73	170	0.68	340	0.33
321	0.51	334	0.14	334	1.66	340	0.66	170	0.31
332	0.49	170	0.12	170	1.57	332	0.64	353	0.30
334	0.49	340	0.12	340	1.56	300	0.60	334	0.25
300	0.42	300	0.09	300	1.33	353	0.59	332	0.24
353	0.40	353	0.09	353	1.33	334	0.58	300	0.21
tot h-t	0.35	tot h-t	0.06	tot h-t	1.11	tot h-t	0.53	tot h-t	0.04
D	0.33	D	0.05	D	1.04	D	0.50	331	0.00
331	0.31	331	0.05	331	1.01	331	0.49	D	0.00

It is also worth noting that, at SLA level, the concentration and inequality indexes of the two benchmark sectors (Textiles and Motor vehicles) are very similar and are about the middle of the industry list. The high-tech sector as a whole is only slightly more clustered that the total manufacturing activity, showing therefore, at this geographical level, no sensible clustering dynamics for the innovative activities.

Table 2.17, where employment is the measured variable, emphasises the geographical polarisation of many high-tech industries in Italy. In particular three high-tech industries (Computer and office machinery, Pharmaceuticals, and Optical and photographic equipment) show a SCR_2 greater than 60%[77].

Table 2.17 Italy spatial concentration and inequality indexes at SLA level (employment)

Sectors	*SCR2*	Sectors	*SNH*	Sectors	*SCV*	Sectors	*LG*	Sectors	*KG*
300	0.75	334	0.27	334	2.32	300	0.80	353	0.52
244	0.68	244	0.26	244	2.28	244	0.77	340	0.44
340	0.65	300	0.25	300	2.25	353	0.75	334	0.42
334	0.64	340	0.23	340	2.16	340	0.74	300	0.40
332	0.59	332	0.22	332	2.11	332	0.73	244	0.39
170	0.52	170	0.16	333	1.78	334	0.72	333	0.30
353	0.49	353	0.15	170	1.77	170	0.71	332	0.27
321	0.48	tot h-t	0.13	353	1.75	321	0.67	170	0.24
tot h-t	0.48	321	0.12	tot h-t	1.60	tot h-t	0.66	321	0.19
331	0.42	331	0.08	321	1.58	331	0.57	tot h-t	0.17
D	0.39	D	0.08	331	1.28	D	0.57	331	0.03
333	0.33	333	0.05	D	1.26	333	0.50	D	0.00

The Linda class of indexes in Italy identifies the metropolitan areas of the greater cities (Milano, Torino, Roma, Napoli, Firenze) as the industry spatial core of many high-tech sectors (as Pharmaceuticals, Computer and office machinery, Aerospace). A different spatial pattern is shown by Electronic components (figure 2.27) in which, apart from the isolated metropolitan areas of Torino, Roma, and Napoli) establishments are clustered in a north-west-ward wedge (with East-west branch ranging from Varese to Padova, and the North-south branch stretching from Milano to Firenze) with a small cluster in the Marche region. The Medical instruments industry (figure 2.28) shows a major concentration in a Northern belt, ranging from Varese to Udine, and a second central cluster composed of Bologna, Forlì and Firenze (and other scattered "metropolitan provinces" such as Torino, Genova, Roma, Napoli, and Bari) The Optical and photographic equipment industry (figure 2.29) is characterised by the prominence of a North-eastern cluster which includes almost all *Province* of the following regions: Trentino-Alto Adige, Veneto, Friuli-Venezia-Giulia (together with Torino, Varese, Milano, Mantova, Firenze, Roma, and Salerno). The Aerospace industry (figure 2.30), as already stressed above, is centred in the metropolitan areas of the largest cities.

[77] Furthermore, Lombardia is within the first 2 *Regioni* (in term of employment) in 6 out of 8 high-tech industries.

Figure 2.12 US Pharmaceuticals spatial core (employment)

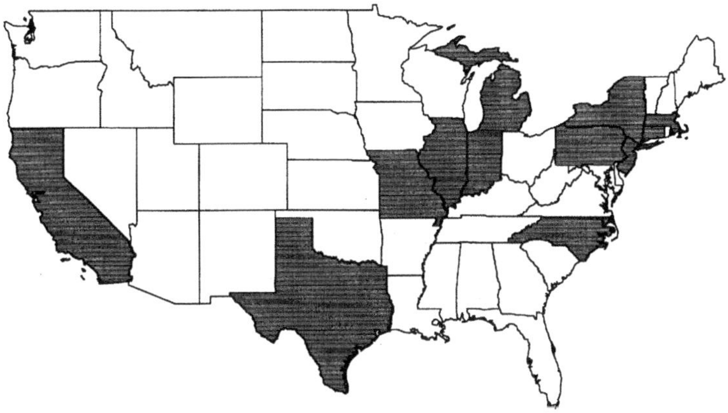

Figure 2.13 US Industrial process control instruments spatial core (employment)

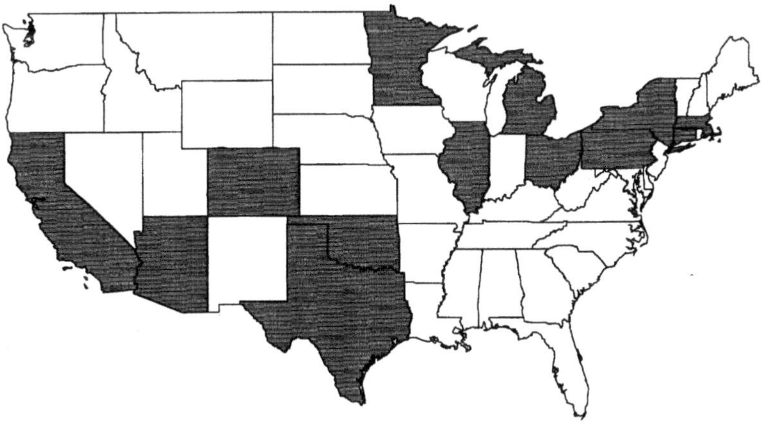

Figure 2.14 US Optical and photo. equipment spatial core (employment)

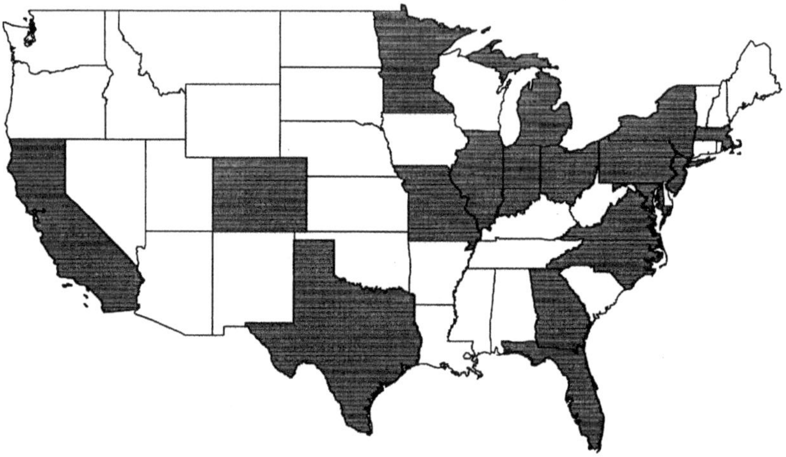

Figure 2.15 US Aerospace spatial core (establishments)

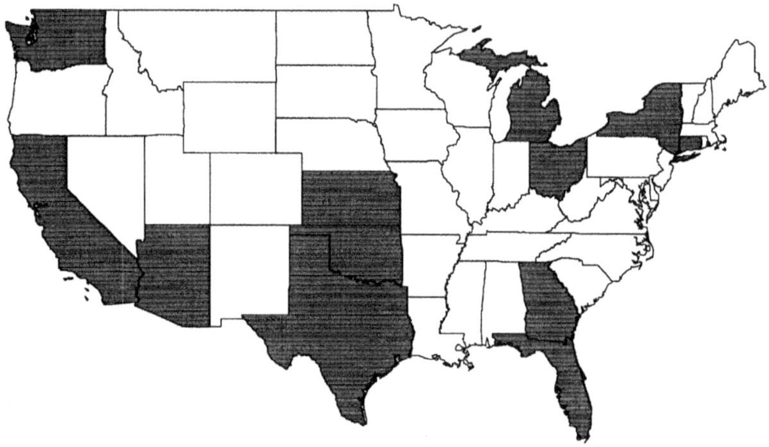

Figure 2.16 US Textiles spatial core (employment)

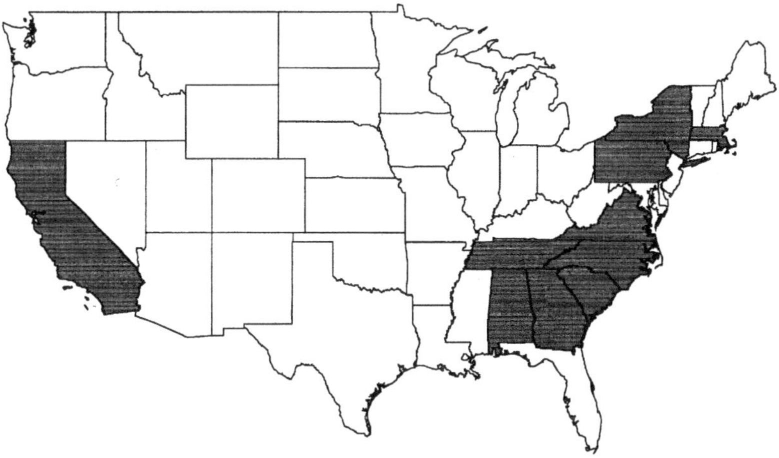

Figure 2.17 US Motor vehicles spatial core (employment)

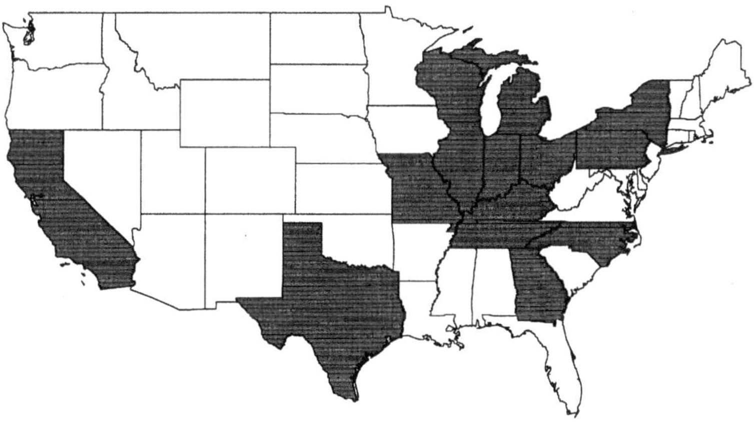

Figure 2.18 UK Pharmaceuticals spatial core (establishments)

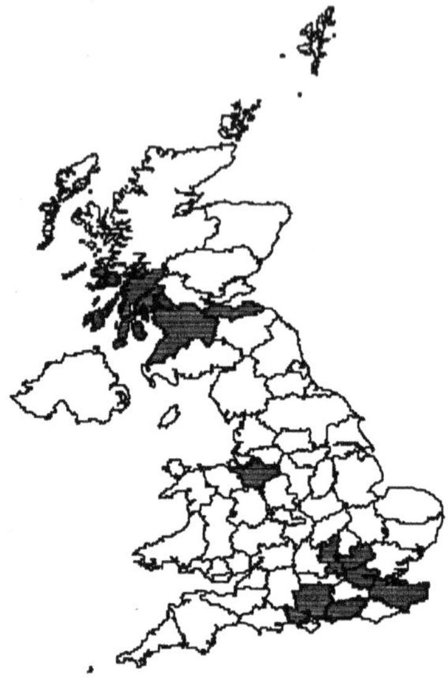

Figure 2.19 UK Computers and office machinery spatial core (establishments)

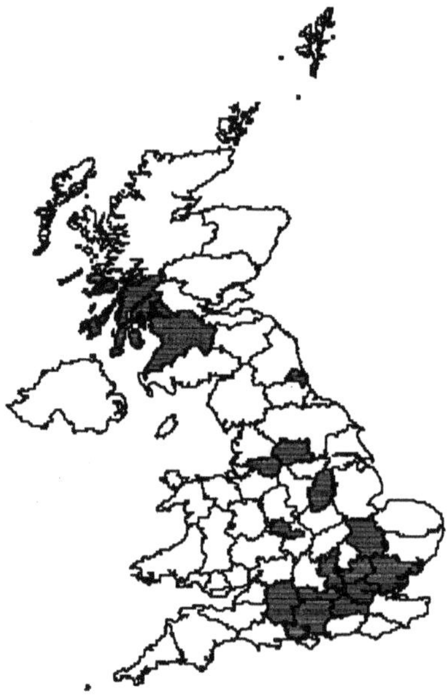

Figure 2.20 UK Electronic components spatial core (employment)

Figure 2.21 UK Industrial process control instrum. spatial core (establishments)

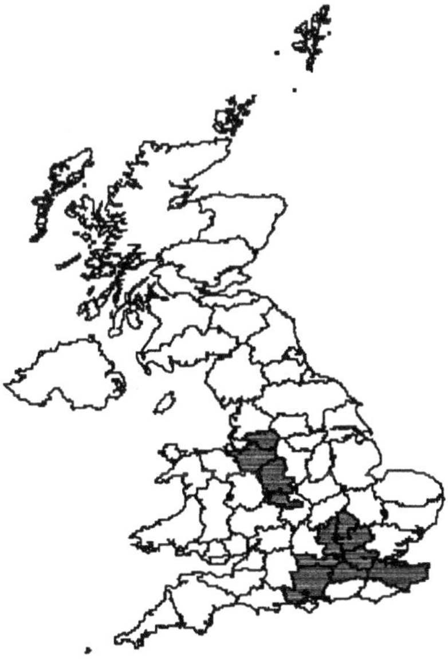

Figure 2.22 UK Optical and photo. equipment spatial core (establishments)

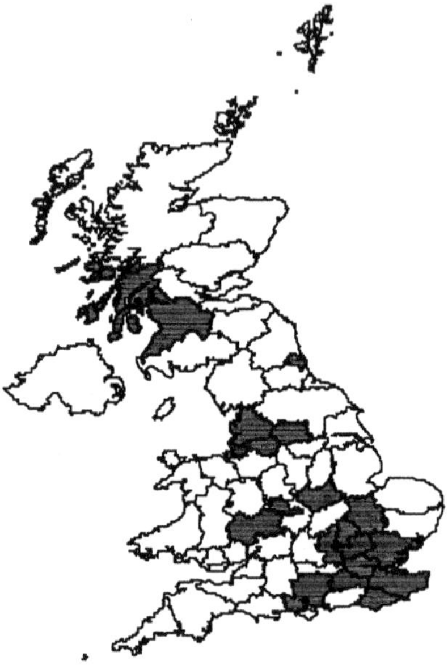

Figure 2.23 UK Aerospace spatial core (employment)

Figure 2.24 France Computers and office machin. spatial core (establishments)

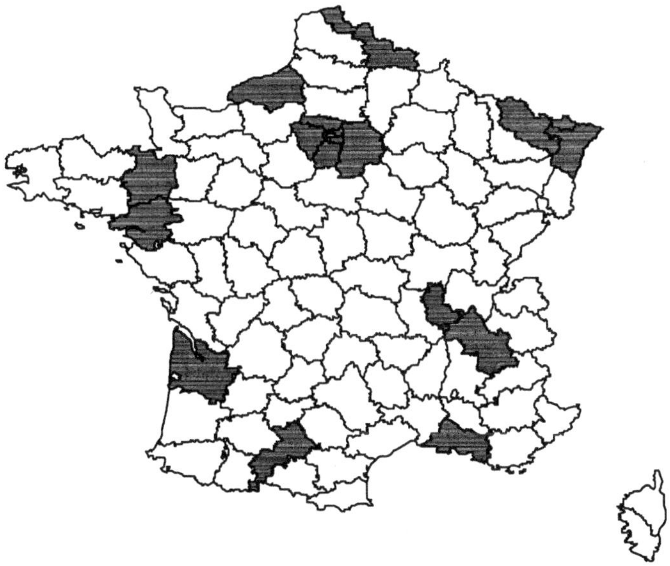

Figure 2.25 France Aerospace spatial core (employment)

Figure 2.26 France Motor Vehicles spatial core (employment)

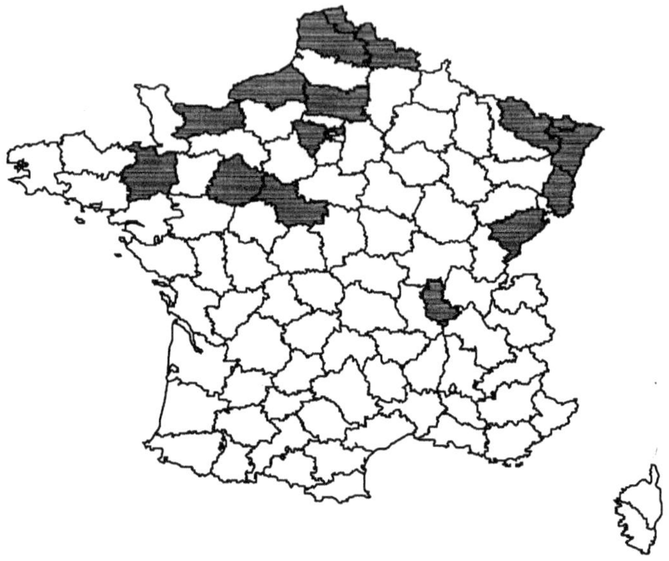

Figure 2.27 Italy Electronic components spatial core (establishments)

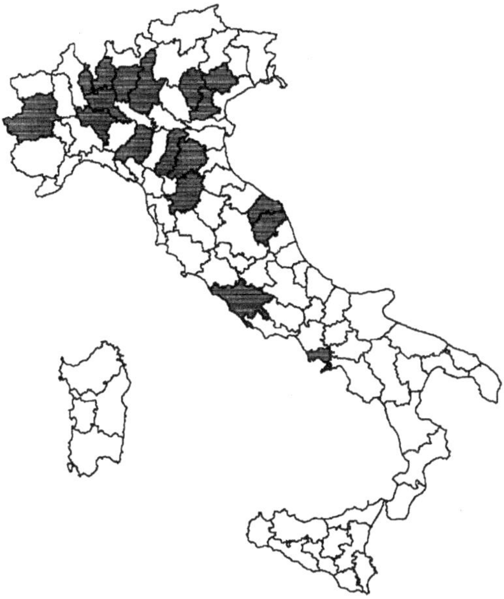

Figure 2.28 Italy Medical and surgical Instruments spatial core (establishments)

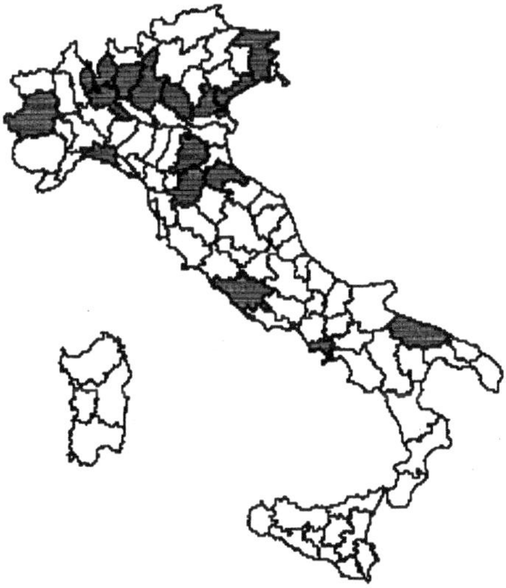

Figure 2.29 Italy Optical and photo. equipment spatial core (employment)

Figure 2.30 Italy Aerospace spatial core (establishments)

iv) International comparison

The first element to be compared across countries — apart from the absolute size of the national high-tech industries, which have been already presented — is the relative specialisation of each nation with respect to high-tech activities as shown in table 2.18.

Table 2.18 High-tech specialisation (total high-tech/total manufacturing)

	USA	UK	France	Italy
establishments	6.06%	8.91%	10.42%	5.00%
employment	12.87%	10.73%	15.33%	5.76%

Contrary to the common wisdom of the US being the most specialised country in high-tech sectors, France appears to have the highest ratios of high-tech establishments and employment[78]. The US follows shortly behind France when employment is considered, but the UK gets into the second position as far as the number of establishments is concerned. Italy is far behind, with a percentage between one third and one half of the leading country.

Another interesting comparison concerns the average size of high-tech establishment in different countries as shown by table 2.19.

Table 2.19 Average size of high-tech establishments (emp./estab.)

	USA	UK	France	Italy
average size	99.27	33.89	80.55	10.16

The data show that while larger plants prevail in the US and in France, smaller establishments are most common in the UK and in Italy. Furthermore the average US high-tech establishment is ten time larger than the average Italian one.

When data on the spatial concentration of high-tech establishments at FLA level are compared, then the US display the highest concentration ratio followed by France, UK and Italy. However it is worth noting that, when this kind of analysis is performed at the single sector level, the most concentrated industry (according to the normalised Herfindhal index) are the Italian Pharmaceuticals[79], followed by US Computers, US electronic components and Italian Optical and photographic equipment.

---

[78] However one must bear in mind that this result may be influenced by the definition of Instruments for France which is wider than those used in the other 3 countries.

[79] Where two metropolitan *Province* (Milano and Roma) record more than 45% of the industry total.

A similar picture emerges when employment data are analysed: Italian Computers and Aerospace take the lead, followed by US Optical and photographic equipment. However one must bear in mind that an SCR_4 of 0.47 for the US Computers industry means more than 106,000 employees, while the Italian SCR_4 of 0.78 refers to less than 20,000 employees. This observation shows clearly that all relative measurements run the risk of overestimating single national and sectoral situations irrespective of their absolute and objective relevance.

It is more interesting to note that, while in Italy and in the US high-tech sectors are far more concentrated at the FLA level, in the UK it is exactly the opposite: all indexes register higher values at the region level than at the county level[80]. For France the situation is a little blurred, since while concentration indexes and the coefficient of variation are higher at the SLA level, Gini coefficients give the opposite results.

Total manufacturing is almost always (at all levels and in all countries) in the bottom part of the concentration lists; however in the UK, at the FLA level, total manufacturing appears to be more concentrated than its high-tech subset.

As far as "traditional" industries (Textiles and Motor vehicles) are concerned, they seem to show the highest concentrated structure in the US and Italy at the SLA level, while in France, at the FLA level, car manufacturing establishments seem to be scattered almost everywhere.

A further international analysis can be performed by comparing the spatial correlation coefficients of different high-tech industries (as in table 2.20).

[80] This result has been interpreted in the following way: it does not really matter in which County an high-tech firm does locate as far as it locates in the South East.

Table 2.20 Spatial association between high-tech industries in four OECD Countries at FLA level (establishments)

USA									*Average*
Industries	244	300	321	331	332	333	334	353	
244		0.33	0.20	0.41	0.35	0.22	0.37	-0.01	*0.48*
300			0.86	0.81	0.82	0.61	0.67	0.25	
321				0.67	0.80	0.65	0.58	0.29	
331					0.70	0.65	0.65	0.21	
332						0.69	0.57	0.20	
333							0.55	0.18	
334								0.15	
353									
UK									*Average*
Industries	244	300	321	331	332	333	334	353	
244		0.24	0.12	0.09	-0.02	0.01	0.16	0.10	*0.27*
300			0.52	0.42	0.64	0.38	0.23	0.23	
321				0.27	0.47	0.18	0.28	0.30	
331					0.43	0.24	0.31	0.37	
332						0.23	0.31	0.35	
333							0.01	0.25	
334								0.31	
353									
France									*Average*
Industries	244	300	321	331	332	333	334	353	
244		-0.16	-0.24	-0.31	-0.20	-0.09	-0.13	-0.37	*0.12*
300			0.24	0.29	-0.02	0.11	0.15	0.44	
321				0.60	0.14	-0.02	0.28	0.67	
331					0.09	-0.09	0.32	0.82	
332						0.08	-0.06	0.55	
333							-0.12	-0.01	
334								0.37	
353									
Italy									*Average*
Industries	244	300	321	331	332	333	334	353	
244		0.35	0.15	0.38	0.50	0.27	-0.06	0.11	*0.20*
300			0.32	0.58	0.47	0.14	-0.08	0.34	
321				0.16	0.36	0.44	0.07	0.06	
331					0.45	0.21	-0.09	0.12	
332						0.48	0.07	-0.02	
333							0.03	-0.04	
334								-0.05	
353									

The table shows significant similarities and dissimilarities across Countries in the spatial distribution of high-tech industries. In the US, Computers and Electronic components display the highest spatial association coefficient, witnessing the existence of an established structure of forward and backward linkages in the information technology sector[81]. In the UK the most spatially associated industries are Computers and Measurement instruments. In France, Aerospace acts as a peculiar industrial attractor, recording high spatial association coefficients with Electronic components and, above all, Medical instruments. In Italy, Computers

[81] The strength of this correlation is further reinforced by noting that the value of the spatial correlation coefficient is the highest in the whole sample.

and Medical instruments are the most significant spatially associated industries. The centrality of the Computer industry in the US is shown by the three highest spatial association coefficients. A similar role, but with a major difference, is played, in the UK, by the Measurement instruments industry. This sector records the two highest coefficients and the lowest one (with Pharmaceuticals). The same happens with Aerospace in France, which records the two highest coefficients and the lowest one (with Pharmaceuticals). Italy shows its weak innovative structure with the lack of any innovative industrial pole. Table 2.19 confirms also the heterogeneity of the Optical and photographic equipment industry within the high-tech sectors in Italy and in France (many spatial association coefficients are negative). The values of the average coefficients show that the US have the most similar geographical distribution for all high-tech industries, followed by UK, Italy[82] and, far below, France[83].

2.3.3 Final remarks and observations

Chapter 2 has been devoted to a geographical analysis of the location of high-tech firms and employment based on an original database — concerning the spatial distribution of 10 high-tech industries within four major OECD countries (US, UK, France, Italy) at two distinct geographical levels (FLAs and SLAs) — which has been expressly built (Maggioni, 1999). Section 2.3.2 has thoroughly discussed the problems which had to be solved in order to make these data comparable. Here, it is worth stressing the contributions that the new empirical evidence presented in this book has brought to the existing knowledge on the clustering of high-tech firms.

Methodology:

1. Absolute and relative measures of clustering give different and complementary information on the spatial distribution of industrial activities, because of the existence of scale economies and threshold effects.
2. Within the toolbox of the economic geographer and of the regional scientist there is a plethora of concentration and inequality indexes which are almost unknown in mainstream economics.
3. Empirical results are crucially dependent on the selected geographical level of analysis, the measured variable, the adopted industrial classification, the chosen concentration and inequality index.

Contents:

1. High-tech activity (measured both in terms of number of establishments and level of employment) shows a significant degree of spatial concentration. Even though clustering is not an exclusive feature of innovative industries, these in-

[82] Where metropolitan areas act as strong attractors for all high-tech industries.

[83] This results may be due to the decentralisation policy implemented by the French government in the Eighties.

dustrial sectors are, in general, more clustered than total manufacturing and the "benchmark" sectors (Textiles and Motor vehicles).

2. US high-tech industry, at the FLA level, displays the highest degree of spatial concentration, followed by French, British and Italian ones.
3. High-tech activities are clustered in the West coast and Mountains States in the US, in the South East Region in the UK, around Paris in France and in the metropolitan areas of the major cities in Italy.
4. In almost all countries, the exception being France, Computer and office machinery displays the highest spatial association coefficients, confirming the relevance of geographical spillovers and the extent of local forward and backward linkages of such industry. In France, Aerospace acts as the most powerful industrial attractor.

Chapter 2 aimed at showing the relevance of firms clustering. The use of several statistical indexes, applied to a number of high-tech sectors in four industrially developed countries, has sufficiently proved that firms do really cluster and that there is plenty of empirical evidence on the agglomeration and clustering of firms (and especially of high-tech ones). This evidence has attracted different streams of theoretical literature which have examined firms' locational choices trying to explain the clustering dynamics and to highlight the role played by different factors. Chapter 3 therefore will review the major theoretical findings in relation to the specific object of this book, i.e. the location of high-tech firms and the emergence of spatially defined industrial cluster.

2.4 Appendix: data sources

This appendix is devoted to the presentation of all data sources used in the book, while the definition of variables can be found in the appropriate chapters.

Chapter 2

Data on the number of establishments and the employment level for 12 sectors[84] in 4 OECD countries (US, UK, France and Italy) at both FLA and SLA level have been collected from, or have expressly been produced by, national official sources.

- United States: *County Business Patterns*, 1994, US Department of Commerce, US Census Bureau, Washington.
- United Kingdom: unpublished 1995 data, expressly produced for the book by ONS.
- France: unpublished 1995 data, expressly produced for the book by INSEE.
- Italy: *Censimento generale dell'industria, commercio, servizi e artigianato 1991*, ISTAT, Roma.

[84] For the US and Italy, 10 for the UK and 9 for France.

Chapter 6

Section 6.2

Data for 51 US States[85] in two different years: 1986 and 1993.

- APAY: Average annual pay (in US dollars), source: *Employment and Wages Annual Averages*, various years, US Bureau of Labour Statistics, Washington.
- ECOL: number of students enrolled in institutions of higher education, source: *Fall Enrolment in Higher Education*, various years, US Department of Education, Washington.
- EMP: Number of mid-march high-tech employees, source: *County Business Patterns*, various years, US Department of Commerce, US Census Bureau, Washington.
- EMPOP: Number of mid-march high-tech employees divided by State resident population, both variables defined as in this paragraph.
- ESTA: Number of high-tech establishments, source: US Department of Commerce, *County Business Patterns*, various years, US Census Bureau, Washington.
- ESTAPOP: Number of high-tech establishments divided by State resident population, both variables defined as in this paragraph.
- EXPORT: State export related shipments as percentage of total shipments, source: *Exports from Manufacturing Establishments*, various years, US Census Bureau, Washington.
- FAIL: Number of business failure by State (in thousand), source: *Business Failure Records*, various years, Dun & Bradstreet Corporation, Wilton.
- HWAY: State highway mileage, source: *Highway Statistics*, various years, US Federal Highway Administration, Washington.
- INCO: Annual personal income per capita, by State (in US dollars) source: *Survey of Current Business*, various years, US Bureau of Economic Analysis, Washington.
- LAND: State Areas, source: *Census of Population*, 1980, US Census Bureau, Washington.
- MALEPT: Ratio of civilian labour force on active population, *Labour Statistics*, various years, US Bureau of Labour Statistics, Washington.
- METRO: Resident population in metropolitan areas as percentage of State population, source: *Census of Population*, various years, US Census Bureau, Washington.
- PAT: Number of patent granted to resident of the United States, source: *Patenting Trends in the US, States and Counties Report*, various years, US Patent and Trademark Office, Washington.

[85] 50 States and the District of Columbia.

- POP: State resident population, source: *Current Population Reports*, various yeas, US Census Bureau, Washington.
- TAX: State marginal corporate income tax rate, source: *US Advisory Commission on Intergovernmental Relations Report*, various years, ACIR, Washington.
- TRD: Total of funds for research and development activities (in millions of US dollars), source: *National Patterns of R&D Resources*, various years, National Science Foundation, Washington.
- UNEMP: Total unemployment as percentage of civilian labour force, source: *Geographic Profile of Employment and Unemployment* , various years, US Bureau of Labour Statistics, Washington.
- UNION: Union membership as percentage of the number of State employees, source: *Manufacturing Climate Study*, various years, Grant/Thorton, Chicago.

Section 6.3

Data on the number of establishments and the employment level for 5 sectors in 4 OECD countries (US, UK, France and Italy) at both FLA and SLA level have been collected from, or have expressly been produced by, national official sources.

- United States: *County Business Patterns*, 1994, US Department of Commerce, US Census Bureau, Washington.
- United Kingdom: unpublished 1995 data, expressly produced for the book by ONS.
- France: unpublished 1995 data, expressly produced for the book by INSEE.
- Italy: *Censimento generale dell'industria, commercio, servizi e artigianato 1991*, ISTAT, Roma.

Section 6.5

The original database contains data on the evolution of 55 high-tech clusters[86] from as early as 1948[87] until 1994. However for expositional reasons, the empirical analysis has been performed on 28 clusters. Clusters are identified by the name of the State followed by the original SIC code (see table 2.1 for details).

- ESTABLISHMENTS: Number of establishments, source: US Department of Commerce, *County Business Patterns*, various years, US Census Bureau, Washington.

---

[86] For each high-tech sector, we chose 8 States which recorded the highest number of establishments either at the initial or at the final date.

[87] County Business Patterns started to record data for different high-tech industries in different years: Instruments and related products were recorded since 1948; Drugs, Computers and office equipment and Aircraft and parts since 1956; Electronic components and accessories since 1959; Guided missiles, space vehicles and parts, and Computer and data processing services since 1974.

Section 6.6

Data on the growth rate of different variables in the period 1975-1995 regarding 112 clusters (state-industries).

- EMPLOYMENT: Number of mid-march employees, source: *County Business Patterns*, various years, US Department of Commerce, US Census Bureau, Washington.
- WAGE: Annual payroll (in US dollars) divided by the number of mid-march employees, source: *County Business Patterns*, various years, US Department of Commerce, US Census Bureau, Washington.

Chapter 3 - Theoretical literature review

> On ne fait point de l'industrie entre ciel et terre; il faut se poser quelque part sur le sol.
> L. Walras (1874), *Elements d'économie politique pure.*

> (...) who can deny the spatial aspects of economic development: that all economic processes exist in space as well as in time? Realistically both time and space must be vital considerations in any theory of economy. Unfortunately, (...) the architects of our finest theoretical structures (...) continue to abstract from the element of space and in doing so they are approaching a position of great imbalance (...) and actually confine themselves to a wonderland of no dimension.
> W. Isard (1949), *The General Theory of Location and Space Economy.*

3.1 Introduction

The two contrasting quotations which open this chapter seem to suggest that the analysis of industrial location (and, more generally, the economic analysis of spatial issues) has been for years a fundamental but, surprisingly, almost neglected topic in mainstream economic theory[1].

The aim of this chapter is therefore to survey different streams of literature and highlight the most significant contributions to the study of the location decision of firms in order to identify the key issues to be first theoretically developed within an original modelling framework and then empirically tested on the already described data set, in the following chapters.

Among the several questions that have been investigated in the literature, three are central: (i) why are there agglomeration (and dispersion) forces; (ii) why are there agglomerations formed by different types of economic agents; (iii) why do clusters specialise in different activities?

---

[1] The situation has notably changed in the early 1990s since the diffusion of Paul Krugman's works which have generated a new wave of theoretical and empirical contributions (which is currently known as "new economic geography").

For this reason the following pages, devoted to a survey of the theoretical literature, are not intended as an exhaustive review of location theory. On the contrary the aim of this chapter is to draw from different streams of literature (and even from different disciplines such as economics, regional science, operational research, corporate geography, business studies) the relevant building blocks for a theoretical approach to explain the emergence of industrial innovative clusters building upon high-tech firms location decisions.

The reader particularly interested in location theory can easily refer to Isard (1956), Hoover (1948), and Beckmann (1968) for a taste of the founding fathers' contributions; to Gabszewicz and Thisse (1992) for a game-theoretical approach; to Stahl (1987) for a detailed survey on urban business locations; to Beckmann and Thisse (1986) for a compact and rigorous summary; to Greenhut and Norman (1995) for an exhaustive and all-comprehensive set of readings; to Ponsard (1983) for further historical details.

The reader more interested in firms' entry and formation processes can see Geroski (1991 and 1993), Acs and Audretsch (1989 and 1992), Storey (1982 and 1985), and the whole issue of *Regional Studies* (1994).

Finally the reader interested in spatial externalities, industrial spillovers and their relations to the process of regional development can see Jaffe (1989), Jaffe et al. (1993), Malecki (1983 1986 and 1991), Malecki and Varaiya (1986), Henderson (1994).

3.2 "Classical" location theory

3.2.1 Least cost approach

The seminal contribution to the theory of industrial location is identified by most scholars in Alfred Weber's book — published in 1909 in German and translated into English twenty years later (Weber 1929) — *Theory of the Location of Industries*[2]. Weber is the founder of the so called "least cost approach" to industrial location. His model explains the location of (manufacturing) firms as arising from the interaction of three factors (three "orientations" in Weber's words): transportation costs, labour costs and agglomeration forces. The first two factors constitute what he called regional (i.e. geographically determined) components whilst the third represents what he defined as non-regional factors (i.e. determined by the interaction of firms' location decisions). Weber's theory is based upon three main assumptions. First, the geographical basis of material inputs is given; second, the situation and size of places of consumption are given (so to imply conditions of

[2] Even though some authors underlines that other scholars (namely von Thünen and Launhardt) had by the time already put the very first building blocks of the discipline some fifty years before.

perfect competition for all producers); third, there are several fixed labour locations with labour immobile and in unlimited supply at fixed wage rates[3].

On these bases Weber constructs a location decision model as follows: firstly firms choose a location (the point of minimum transport costs) which would allow the minimisation of transport costs between the sources of material inputs and the market for output (through the construction of a locational figure[4]), then they take into account two distorting forces which can shift the firms' location from the previously calculated optimum site: the location of cheap labour and the effect of agglomerative (and deglomerative) forces.

Weber takes into account the existence of points of lower labour costs which can make the firm deviate — in any direction — from the point of minimum transport costs. By drawing a series of lines (isodipanes) centred around the point of minimum transport costs, it is thus possible to identify all locations for which additional transportation costs are equal to a given series of values. A "critical" isodipane is then built as the *locus* of points for which the saving in labour costs is equal to the extra cost of increased transportation to the new location (per ton of product). The location of the firms will therefore deviate towards the point of lower labour costs if, and only if, this point is located inside the area delimited by the critical isodipane[5].

---

[3] Wages, and more importantly, total labour costs — which take into account differences in "subjective efficiency" (a function of a particular geographic distribution of the population) — can be different at different locations and, given the hypothesis of immobility and unlimited supply of labour force, stay different forever. No labour migration is in fact allowed in the model.

[4] In the most famous example a firm is producing a good by using two inputs, available at two different locations, and sells its output on a market, situated in a third location. The location figure — used to calculate the point of minimum transport costs — is therefore a triangle (known as "location triangle"). The actual location point is then determined by the force (a function of the "ton per miles" coefficient that can be saved by approaching the corresponding corner) with which each corner attracts the firm. The ratio between the weight of localised raw materials and the weight of finished product (which Weber calls "material index") determines the first approximation of the firm's location. If the index is greater than one, the firm will locate towards the source of raw material; if it is smaller than the unity, the firm will locate towards the market place. On the basis of these intuitions the empirical literature has tried to build an exhaustive taxonomy of industries with respect to their locational decisions: materials oriented are those industries (as Chemicals and Metal products) which are primarily determined by the location of inputs, market oriented are those industries (as Food and beverage) heavily influenced by the spatial distribution of consumers, and footloose industries (sometimes identified with high-tech ones) are those industries whose location dynamics are scarcely influenced by the spatial distribution of both inputs and consumers.

[5] Similarly to material index, a "coefficient of labour" — calculated as the ratio between the index of labour and the "locational weight" — measures the force of attraction of labour. High coefficients of labour will lead to concentration of firms around the cheap labour source; low coefficients, to their dispersion.

The third set of factors influencing the firm's location decisions are the economies (both positive and negative) arising from the spatial clustering of firms[6]. Agglomeration will arise if the isodipanes of a sufficient number of firms intersect and if the total output of these firms is greater than or equal to the quantity required to benefit from the economies of agglomeration. Actual firms clustering will emerge in the intersection area of several firms critical isodipanes. The actual ordering of three criteria (first transport, then labour, finally agglomeration) is more a logical than an historical one. In theory, firms firstly take into account the spatial distribution of relevant inputs and output (remember labour is an ubiquitous resource), then search locally for cheaper labour; finally they consider (although not in a strategic way) the effects of other firms location.

Other authors such as Palander (1935), Hoover (1937 and 1948) and Isard (1949 and 1956) further contributed to the development of this approach to industrial location by enlarging its scope and enforcing its predictive power.

Palander's thesis was published in 1935. In this early work he addresses two fundamental questions. First — given the location of materials and the position of products market — where will location take place[7]? Second — given the location of firms, competitive conditions, production and transportation costs — how does price affect the extent of the area in which the firm can sell its output? The contribution of Palander (1935), thus allows the least cost approach to deal with the issues of spatially distributed consumers and market areas[8], to clearly distinguish in the analysis of total costs between production and transportation costs, to analyse the effects on isodipanes arising from non-linear freight rates, and to integrate the locational problem into more general production theory.

Later contributions in the same stream pursued the connection between location and production theory in several ways: Hoover (1937 and 1948) investigates the effects of diminishing returns in production on market area boundaries; Isard (1951 and 1956) and Moses (1958) show that location decisions can be analysed within the traditional framework of the substitution principle between factors[9]. In

[6] In particular Weber assumes the agglomeration force to be a parabolic (quadratic, concave) function of the number of firms and the "deglomeration" force to be linear. Once again an index of saving in unit costs, dependent on agglomeration, is defined as the reduction of production and marketing costs resulting from firms clustering.

[7] This first question was very much in the tradition of Weber, even though Palander blamed Weber's analysis of agglomeration economies because of the lack of strategic considerations and of a dynamic perspective.

[8] Even if this is done in a very specific way: costs are still the only relevant variable and demand is held constant.

[9] The aim of the substitution approach to location theory is efficaciously summarised by Greenhut and Moses. "The theory of plant location is one segment of economic theory. It, too, rests on the principle of substitution. The extent to which labour can be substituted for capital or land and vice versa is basically the same problem as the selection of a plant site from among alternative locations. Both decisions attempt to maximize the ends. The objective is accomplished when the scarce means are allocated among competing ends in the optimum manner." (Greenhut 1956, p. 4). "My objective is to place the theory of location within the main body of economic literature. I wish to make the theory of location an integral part of the theory of production and to investigate the implication of

Moses (1958) the optimal location can therefore be calculated by identifying the tangency point between the higher isoquant (Isard's "transformation line") and the iso-outlay line (Isard's "equal outlay line"), where the ratio of the marginal productivity of spatially located production factors is equal to the ratio of their delivered price. Moses extends the analysis even further by highlighting the fact that, if the production function is not homogeneous of the first degree, then there is no single optimum location, for the optimum location varies with the level of output[10].

The above mentioned authors made a very substantial point in highlighting that the concept of perfectly competitive general equilibrium (PCGE) — à la Walras, Cassel, Pareto — cannot be meaningfully applied to spatial economics[11]. A spatial PCGE framework would necessarily imply the following highly unrealistic assumptions: transport costs equal to zero; capital and labour ubiquitous or perfectly mobile; production technology uniform throughout the space.

These authors concentrated therefore on partial equilibrium analysis, focusing on the supply side of the economy; the driving force of location decisions being the minimisation of costs, allowing little, if any, room for demand variation and strategic interaction between firms[12] (Beckmann 1990).

Although at a first glance these contributions may seem rather primitive and very distant from the topic at study — transport costs being almost negligible for high-tech products[13] (whose typical weight to value ratio is very low) — it must be stressed that they can handle very effectively some peculiar features of innovative sectors.

The first feature lies in the understatement of the demand side of the locational issue, related to the geographical dispersion of consumers, and in the consequent

---

factor substitution for the locational equilibrium of the firms. My main conclusion is that profit maximization requires a proper adjustment of output-input combination, location and price. Moreover, the optimizing values of these three variables can be determined with analytical tools derived directly from traditional economic theory. There is no need for much of the esoteric paraphernalia sometimes employed by location specialists." (Moses 1958, p. 259).

[10] By introducing the possibility of a plurality of possible equilibria, Moses directly calls for the introduction of "demand considerations" to identify the actual equilibrium. However this procedure is not explicitly and analytically dealt in his analysis.

[11] The same point has been recently recognised by Krugman (1991a and 1995) as the main cause for the exclusion of spatial issues from the realm of economic theory.

[12] The existence and the exploitation of agglomeration economies for an individual firm in Weber's model depends strongly on the behaviour of other firms, even though these behaviours are assumed as given. It must be noted, however, that this remark does not apply to some more recent contributors of the least cost approach. From the late Sixties, in fact, Isard (Isard 1966; Isard and Smith 1967) has repeatedly stressed the importance of game theory as an useful tool to model locational problems and to understand locational patterns subject to agglomeration dynamics.

[13] It must be stressed, however, that a re-interpretation of transport costs as distance-related opportunity costs, would radically change the picture. Closeness to intermediate input producers is very important for high-tech firms because of the relevance of producer-user interactions in the innovative activity (von Hippel 1988).

neglect of market areas. Innovative firms in real economic life either produce very specific consumer products whose relevant market extend to the "whole world" (therefore there are no locational advantages connected to any specific site to serve such a market) or they produce some intermediate goods used as input by other firms, which are usually larger and spatially concentrated.

The second feature concerns price determination. Weber's hypothesis of perfect competition does not sound so unrealistic in the specific sectoral framework. In some high-tech industries, firms seems to have very little power in fixing prices: on the one hand, the world market is so large — compared with the output potential of a single firm — that it seems rather sensible to model the situation as an almost perfect competitive environment where firms are price takers; on the other hand, when firms are subcontractors or suppliers of larger ones, then the situation resembles an oligopsony where few buyers fix the price and sellers are price taker.

The third feature regards the relevance of (cheap) labour force location. High-tech firms do require skilled workers and it is rather simple to show that — if one considers the hedonic (or quality-adjusted) price of labour, by computing both education and training expenses as part of the labour costs — the search for particular locations which are endowed with a richer pool of (already trained) workers does correspond to the original orientation towards cheap labour locations in the least cost approach. It must be also considered that these pools of skilled labour are often produced by already existing clusters of similar firms. It is thus rather difficult to distinguish labour orientation from the willingness to experience agglomeration economies. In this case the new entrant firm can free-ride (at least in the short run) on the local labour pool and "poach" skilled workers (by offering marginally higher wages) who have been already trained by other firms in the area[14].

The fourth feature deals with agglomeration economies. Especially in high-tech sectors, phenomena such as: information spillovers, transaction costs reductions, existence of a pool of specialised service firms and spin-offs have a crucial influence in determining the location of industry (as extensively documented in the second chapter). Therefore such an approach, stressing the relevance of agglomeration forces and processes in determining the location of economic activities, can be useful in interpreting the empirical sectoral evidence.

The fifth feature refers to the existence of multiple locational equilibria arising from the substitutability of inputs and on the condition for the production function to exhibit increasing returns to scale. This original result, due to Moses (1958), solves a possible conflict between reality and the Weberian location doctrine (stating that there is only one single best location for each type of activity) and sheds light on the fact that reality presents numerous examples in which firms in a given industry have different orientation. A key element of high-tech clusters is in fact the plurality of these experiences in the same country.

[14] For an interesting analysis of the relationships between training costs, labour market dynamics, and innovative activity, see Acemoglu (1992).

3.2.2 "Demand side" approach

A complementary perspective on locational dynamics arises from the consideration of the demand side. If buyers are not concentrated in few market locations — which can be served (in spite of differences in profits) by all producers — but are dispersed in a wide area, and costs are assumed constant; then producers will seek the location granting them the greatest sales volume. In this way, by attempting to supply the largest possible market area, they will maximise their revenues and consequently (by the previous assumption of constant costs) they will maximise profits too.

In a sense, models belonging to this group can be considered as a sort of dual of the "least cost" approach, presented in the previous section. There, demand is given and the firm's optimisation problem is reduced to the minimisation of costs; here costs are given and the problem to be solved concerns the maximisation of revenues, subject to the constraint imposed by the decisions and actions of competitors.

The theories belonging to the demand side approach have been alternatively labelled as "market area" or "locational interdependence", according to the relative weights attributed either (i) to the identification of the exact shape and extension of the influence zone of different markets and producers (Fetter 1924); or (ii) to the strategic location decisions that a firm has to take in order to achieve a greater monopolistic (or quasi-monopolistic) power over a number of consumers dispersed within a given area (Hotelling 1929).

An enormous stream of literature[15] has originated from these contributions (and especially from the latter) spelling out all the different hypotheses which determines different patterns of spatial location. However, for the problem at hand, it is not necessary to go into many details, which can be better dealt with in section 3.4, devoted to game theory.

Perhaps it is worth recalling here the contribution of Lösch (originally published in German in 1940 and subsequently translated into English in 1954) because it represents a sort of bridge between the "least cost" and the "demand side" approach and it is an anticipation of the general equilibrium analysis of location.

Spatial equilibrium in this model is determined by two fundamental tendencies: the maximisation of individual advantages (firm's profits) and the maximisation of the number of independent economic units (firms). The equilibrium solution(s) must satisfy five crucial conditions: (i) the location of each firm must be as advantageous as possible; (ii) firms' locations must be numerous enough to cover the entire space; (iii) regions and/or situation of abnormal profits must disappear; (iv) production and supply areas must be as small as possible; (v) borderlines between two market areas must be iso-price lines for consumers.

Even though Lösch does not explicitly calculate the analytical solution of the location model, he states that, in general, the optimal location for industrial production (and consumption) is clustered, while optimal locations for agricultural production are uniformly distributed. Furthermore Lösch explicitly states that lo-

[15] Summarised in Dean and McKee (1970), analytically reformulated in Gabszewicz and Thisse (1986), and extensively well documented in Greenhut and Norman (1995).

cational interdependence approach — when confronted with equilibrium requirements — is doomed to produce analytical solutions cursed by mathematical impossibility and that the individual firm, endowed with a limited rationality, will often end up in sub-optimal locations[16].

The main limitation of the demand side approach lies in the partiality of the representation of inter-firm relationships, which are conceived only as competitive, leaving aside any other possible interaction (such as technical complementarities, market synergies, R&D co-operation etc.).

It must however be emphasised that this approach has firstly shown that spatial clustering of firms can be caused by factors other than agglomeration economies (as in the principle of minimum differentiation presented in Hotelling 1929).

Subsequent contributions, such as Eaton and Lipsey (1976) and many others, have then added further theoretical explanations for demand-driven firms' clustering, such as "comparative shopping" and indivisibility of the buyer. Although originally designed to describe household behaviour, these two reasons can be fruitfully applied to analyse the location process of high-tech firms. It seems reasonable to assume (following the "comparative shopping" line of reasoning) that an industrial customer — which has to buy intermediate inputs from a number of small high-tech firms — would prefer to do business with firms located in clusters (where there are several firms producing similar or at least comparable products) rather than with geographically dispersed firms. By a similar reasoning, the principle of indivisibility of the buyer can also explain why clusters of firms specialising in related or complementary intermediate products are more attractive to industrial buyers.

3.2.3 "Land utilisation" approach

This approach underlines that space plays an absolutely unique role in economic theory. On the one hand (as widely illustrated by the previous approaches) space is something which causes disutility, by imposing transportation and communication costs to bridge distances; on the other hand, it is a truly economic good — which classical economists called "land" — which provides utility as a necessary and generally scarce production input and which has its own price: rent.

The merit of originating this approach belongs to von Thünen who, in 1875, developed a model in which the location patterns of productive activities (around a city centre) are dependent on different land costs. More profitable activities achieve better (i.e. more central) locations by being able to pay higher rents; less profitable ones are relegated to the outskirts of the city where rents are lower.

---

[16] "If we wish to be precise and to consider the influence of the selection of a particular location on all other locations (...) (then) we are faced with the interdependence of all locations. Equilibrium of the location system can therefore no longer be charted, but can be represented only by a system of equations that are insoluble in practice. (...) There is no scientific and unequivocal solution for the location of the individual firm; but only a practical one: the test of trial and error." (Lösch 1954, p. 8 and p. 29).

It must be stressed, however, that some of these conclusions depend heavily on land market imperfections. In the presence of a perfectly clearing market for land, all locational benefits (included those arising from agglomeration economies) would be capitalised in land values. If this is the case, then the entire geographic space, in equilibrium, should display a perfect locational indifference; for each economic agent, wherever located, the algebraic sum of locational advantages and disadvantages (inclusive of rent paid to the land owner) should be the same[17]. In this situation, all locations are — from the economist's point of view — identical and the choice of location would be of no interest, having no effect on firms' performance and profitability.

Despite the fact that this model (that in certain respects could have been grouped in the least cost approach[18]) was originally designed to describe the locational pattern of different agricultural productions, its validity also holds for more advanced sectors and production. Real life high-tech clusters often find crucial barriers to their expansion in increased land prices both for industrial and residential use, and better positions in these clusters are always occupied by firms belonging to higher value added sub-sectors (as an example, one can compare the location patterns of software houses and "chip" manufacturers within the Silicon Valley).

3.3 General location equilibrium models and urban economic theory

Since the very beginning of locational analysis, several authors have tried, from different perspectives, to move from a partial equilibrium approach to a general equilibrium representation of the location decision. The ambitious goal was to determine the simultaneous location equilibrium of all agents in a given economy.

However this has proved to be a difficult task. For a model to be realistic and interesting, it must display some sort of indivisibilities; but the very presence of indivisibilities may cause the non existence of a locational equilibrium[19]. The so-

---

[17] To be precise, this reasoning will hold only in a "least costs" modelling framework. In the demand side approach, the value of an industrial site depends critically on how many neighbouring sites are already in use as inputs in the same industry, and on customers' location.

[18] Being the rent level dependent on the distance from the market and not, as in Ricardo, from the difference in fertility of the land.

[19] "If all sites are homogenous and all plants are indivisible, and if the operation of plants requires them to exchange their outputs with each other, then no feasible location assignment can be sustained by a competitive price system". (Fujita 1990, p. 185). This derives from a more general result known as the "spatial impossibility theorem" (Starret 1978), which assumes an economy with no relocation costs, homogeneous space (i.e. evenly distribution of immobile resources) and perfect markets for all gods at all locations. Even in such a situation, there is no competitive equilibrium where total transport cost are positive (i.e. the only feasible competitive equilibrium is complete autarky: a

lution has thus been found, in most cases, by assuming that all production activities are perfectly divisible[20].

According to Fujita (1990) we can usefully divide this class of models into four main groups: comparative advantage, non-price interactions, monopolistic competition and oligopolistic interaction. In particular, this section will mainly deal with the two central classes which seem to fit better the case of high-tech firms[21]. In both classes (non price interactions and monopolistic competition) firms' interactions endogenously generate agglomeration forces. This happens because some agents use spatial distribution signals (i.e. information about the spatial distribution of agents) in determining their optimal choice.

3.3.1 Non-price interactions models

Non-price interaction models describe competitive spatial equilibria allowing two different sources of interactions among firms and/or households: a generic locational choice (spatial externality) and a specific activity (communication). In both cases the value of the output produced by a firm is a function of the amount of inputs employed by the firm, but also depends on the interaction with other firms (i.e. there are external economies of scale).

However, the former type (Fujita 1985; Henderson 1977; Kanemoto 1980) uses the concept of "accessibility measure" to implicitly represent the effects of non price interactions and assumes that firms in the same area receive the same amount of external economies regardless of their positioning and active behaviours. The latter approach (extensively illustrated in this section) requires firms to engage in a communication activity and explicitly models agglomeration economies as arising from flows of communications among firms exchanging information[22].

In these contributions, information is modelled as an imperfect public good. The use of a piece of information by a firm does not reduce the content (even though it may reduce the value) of that information for any other firms. Hence the exchange of information, through communication activities, generates externality-

situation where no good or person is transported, production and consumption must necessarily coincide).

[20] To avoid the triviality of a uniform spatial allocation of plants it is also generally required all the sites to be not self-sufficient.

[21] We are therefore explicitly omitting to consider competitive advantage models, since the uneven distribution of immobile resources has been already dealt in the least cost approach and it does not seem to be relevant for high-tech industries. We also do not focus on oligopolistic interaction models with discrete agents, because the case of large firms interacting oligopolistically (i.e. taking into account their perceived market power) through land and labour markets with possible exchanges of intermediate goods does not seem to describe the typical inter-firms relationships established within high-tech clusters.

[22] It must be noted, however, that "these two forms of non price interactions models are mathematically equivalent" (Fujita 1990, p. 190).

like benefits to each firm. These benefits increase geometrically as the number of firms increases linearly[23] and they are greater if firms locate close to each others, because communication involves distance sensitive costs. The final results of this process is thus the clustering of firms in order to ease the flow of communication (Fujita and Smith 1990).

Different authors describe in detail different forms and contents of inter-firm relationships. Tauchen and Witte (1983) explicitly consider face-to-face interactions between similar firms subject to transportation and facility costs. Even though this model was originally formulated to describe the location decisions of offices within central business districts, in my opinion it may also be used to describe the spatial clustering of high-tech firms due to the enormous relevance that knowledge interactions and information spillovers have in advanced industries. In the model, agglomeration economies (which derive from a greater accessibility to contact at more central locations) are balanced by higher land rent and by the intrinsic concavity of the contacts benefits function[24].

Goldstein and Gronberg (1984) model the interactions between complementary firms in a (metropolitan) area as a way — alternative to integration (both vertical and horizontal) — to enjoy economies of scope[25]. They are able to demonstrate, among other things, that "there are economies of agglomeration if firms can make use of a "sharable factor" at a specific location" (Goldstein and Gronberg 1984, p. 102). This approach seems thus to perfectly fit the typical inter-firm relationship existing in structured groups of firms such as the Italian industrial districts and the US high-tech clusters, where technological and marketing complementarities and externalities allow for vertical and horizontal disintegration.

Pascall and McCall (1980) — by underlining the conditions of uncertainty and imperfect information inherent to locational decisions — analyse the relationship between already located firms and outsider firms which are still deciding what location to choose. "It seems clear that (in presence of imperfect information) firms' decision to enter a particular industry should be based on the observed behaviour of the firms in that industry. If location is important then this should be reflected in the locational choices of firms already in the industry. (...) Indeed the number of successful firms in this industry that are located at a particular site indicates the productivity of the site" (Pascal and McCall 1980, p. 384). This paper thus gives a

---

[23] The number of directed communication channels in a n-agents set is in fact equal to $n(n-1)$.

[24] "The net revenue (revenue after payment of all costs other than contacts costs and rent) of a firm depends on the number of contacts it has with other firms in the central business districts. (...) The function is non negative and increasing; it may be eventually decreasing and even negative. This means that there are positive benefits to some level of contact with any other firm, but that the value of additional contacts with any one firm eventually diminishes. The function is (thus) strictly concave throughout" (Tauchen and Witte 1983, p. 1313).

[25] "Economies of scope with respect to space is our concept of agglomerative economies. Agglomerative economies exist when it is less costly to combine two or more product lines in one (urban) area (but not necessarily one firm) than to produce them in different (urban) areas" (Goldstein and Gronberg 1984, p. 97).

theoretical foundation to the empirical evidence, provided by surveys and case studies, which indicate that previous locations of similar firms (i.e. belonging to the same industry and dimensional class) act as a strong signal of the profitability of doing business in the area and therefore positively influences locational choices.

All these models — which stress the relevance of inter-firms relationships (such as knowledge spillovers and informational exchanges) in determining the successful development of an industrial cluster and highlight the signalling function of previous locations — indirectly provide a micro-economic foundation for the modelling framework put forward in chapter four of this thesis.

3.3.2 Monopolistic competition models

Monopolistic competition models assume that price interactions (i.e. exchanges of goods and services through the mediation of prices) are the fundamental causes of spatial agglomeration of economic activities. These models[26] are thus confirming the empirical evidence, frequently quoted in the surveys, that "one of the major causes of industrial agglomeration is the availability of specialised local producer services" (Fujita 1990, p. 198).

Rivera-Batiz (1988) specifically focuses on the service sector. In particular, he shows that the local availability of specialised business services (i.e. transport, maintenance, financial, legal, fiscal, marketing and technological consultancy etc.) can generate agglomeration economies through the following causal chain. "The agglomeration of producers in a particular industry will result in gains if the extent of the market allows producer services to proliferate and to become more specialised, raising the productivity of the industry that use them" (Rivera-Batiz 1988, p. 126).

Fujita (1990) models a similar situation in which a cluster (in his words, a city) is characterised by two industries: an export-good industry (e-industry) and a service-good industry (s-industry) and explicitly states that the e-industry "can be interpreted as the headquarters of high-technology firms (e.g. Silicon Valley)" (Fujita 1990, p. 199). In this model "given the spatial distribution of e-firms, each s-firm chooses its optimal location and f.o.b. price of its s-good. (...) In turn, given the spatial distribution of s-firms, each e-firm chooses its optimal location and consumption level of each s-good. Since s-firms prefer to locate near e-firms and vice versa, we can expect that in equilibrium all firms will form a cluster. Notice also that as the number of s-goods (provided in the city) increases, each e-firm can enjoy a higher productivity. Therefore, this model can also explain the formation of a specialised city due to the availability of specialised local producer services" (Fujita 1990, p. 199).

---

[26] It must be stressed that, even though his models — because of their qualitative and quantitative relevance — have been surveyed under a specific heading, most of the contribution recently produced by Krugman on "geographical" issues may well be ordered under this section.

Kanemoto (1990) demonstrates that market transactions of intermediate inputs (one of which can be information) can create firm clustering if they are coupled with indivisibility (or, more generally, scale economy) in production[27]. This result thus provides a theoretical foundation for those agglomeration economies arising from technological and productive interactions which have been often reported by empirical studies of high-tech clusters (such as Saxenian 1994).

Other contributions explicitly model interactions between firms' and households' location decisions[28] to explain the existence of (mainly urban) agglomeration economies. Papageorgiou (1979) establishes an analogy between vertical, horizontal integration and urban agglomeration. Agglomeration advantages refer to production and consumption while disadvantages refer to transportation costs. The city is thus conceived as a centre of production, a public good and a service centre. Especially the third function of the city provides fuel for agglomeration economies even when there are no output externalities.

Papageorgiou and Thisse (1985), by using a linear city model, stress the interaction between firms and households[29] in the search for an optimum urban dimension. "Households are attracted to places where the density of firms is high because opportunities are more numerous; and they are repulsed by places where the density of households is high because they dislike congestion. Firms are attracted to places where the density of consumers is high because the expected volume of business is large and they are repulsed by places where the density of sellers is higher because of the stronger competition prevailing there" (Papageorgiou and Thisse 1985, p. 20). In this model, the clustering of households and firms requires the existence of boundaries and spatial interaction; firms further require the existence of positive profits, i.e., some kind of imperfect competition. These results can therefore be interpreted — although in a very broad sense — as a theoretical explanation for the positioning of high-tech clusters (especially in the US) close, but not too close, to large cities, due to the interaction of the requirements of high-tech firms and highly skilled employees.

Fujita (1988) models a situation where there is a continuum of firms, each producing a single and distinct commodity, and a continuum of homogeneous house-

---

[27] "In this paper, therefore, interactions between firms are market transactions and do not by themselves represent externalities. Combining the market exchange of intermediate inputs with indivisibility, however, creates externalities in locational decisions. For example suppose that two firms interact with each other and they equally share the interaction costs. If one firm moves closer to the other firm, the interaction costs for both firms decrease. The firm which moves, therefore, gives external benefits to the other firm" (Kanemoto 1990, p. 47).

[28] "In our model firms prefer to locate near households and vice versa. Other things being equal, the centre of the distribution of households is the best location for each firm; the centre of the distribution of firms is the best location for each household" (Fujita 1988).

[29] "Our model rest upon the idea of a complete spatial interdependence between firms and households. The optimal behaviour of a firm depends on what households and other firms do, while the optimal behaviour of a household depends on what firms and other household do: space here binds everyone together" (Papageorgiou and Thisse 1985, p. 29).

holds. Firms are assumed to prefer locations near to households and vice versa. Therefore the centre of the distribution of households is the best location for each firm and *vice versa.* When the number of households is larger than that of firms, we can expect that, in equilibrium, the firms will form a cluster in the centre of the distribution of households. This model thus explicitly explains the formation of downtown or commercial areas; however, "with appropriate modification (...) it will be also possible to explain various type of spatial agglomeration such as business firm agglomeration and high-tech firm agglomeration" (Fujita 1988, p. 89).

Section 3.3.1 and 3.3.2 thus have shown that economists have thoroughly analysed the concept of agglomeration economies and have been able to identify a number of different theoretical reasons which can even explain the clustering of footloose activities such as high-tech industry.

3.4 Game theory

The relevance of firms' strategic behaviour in the choice of locations has been raised in the location analysis literature since the works of Hotelling (1929) and Palander (1935) and further emphasised by Isard (1966). However, only since the 80s, has this issue become the central focus of analysis of a series of contributions, mainly derived from Hotelling (1929), aimed at analysing the strategic side of location choice. "Space, by its very nature, is a source of market power. (...) Because market activities are performed at dispersed points in space, each firm finds only a few rivals in its immediate neighbourhood. (...) Competition in space occurs 'among the few', thus leading to an analysis of the problem as a game of strategy" (Gabszewicz and Thisse 1992, p. 282).

According to Gabszewicz and Thisse (1992), who have extensively reviewed and acutely organised the literature on the strategic approach to location analysis, the locational problem is modelled as a typical noncooperative game in which firms are the players, prices and/or location are the strategies, and profit functions are the payoffs. Location, for game theory, is essentially a matter of spatial competition. Therefore, within this theoretical framework, different models have been proposed in order to deal with different forms of competition, such as location under mill price competition, location under discriminatory pricing, location under non-price competition.

It is very difficult to summarise such a literature which, by its very nature, underlines the weakness of any general theory and would oblige one to deal with a plethora of specific hypotheses. For the purposes of the thesis, however, it seems more relevant to briefly describe advantages and drawbacks of this theoretical approach when dealing with the location of high-tech firms.

Undoubtedly, game theory allows a more precise analysis of spatial competition by underlining the role played by few, sometimes hidden, hypotheses which are crucial for the solution of most locational games (is location sequential or simultaneous? Is there any co-ordination device or not?). In a sense, one may state that the best contribution of this class of models has been achieved in showing the inconsistency, or the severe limitations, of previously considered general principles and laws.

However this approach suffers from some drawbacks by having confined its attention to competitive interactions between firms, leaving aside the role played by technological and informational synergies and externalities in determining actual firms' locations. Further, these competitive interactions are almost always modelled as non co-operative games, avoiding the analysis of the behaviour of groups and coalitions of agents (which could be of extreme interest for the analysis of birth and development of high-tech clusters and science parks). Finally, these models often underestimate the role played by transport infrastructures in influencing firms location decisions. However, some steps aimed at overcome these limitations are presented in the following papers, which study alternative formulations of the original model of Hotelling.

Scotchmer and Thisse (1992) highlight the relevance of spatial issues in the realm of public economics, and show that, once space is introduced, the line between private and public goods is blurred. Furthermore they illustrate that the original result of minimum spatial differentiation (or clustering), obtained by Hotelling and confuted by d'Aspremont et al. (1979), holds as far as the preferences of consumers are sufficiently dispersed[30]. Finally they show that models derived from Hotelling (1929) ignore an important consequence of space, namely capitalisation[31]. Bester et al. (1991) demonstrate that Hotelling's model, with quadratic consumer transport costs possesses an infinity of equilibria which are normally ruled out by imposing a co-ordination device, concerning the ranking of the firms' locations along the market segment and restricting the firms' strategy spaces. Knoblauch (1991) reformulates the same location game of two firms on different finite graphs, obtaining surprising results such as the absence or presence of pure Nash equilibria depending on the topological quality of the graph. This last result is particularly interesting when one wants to assess the empirical relevance of theoretical models in real life situation, where space is not an undifferentiated plain and transport infrastructures are present and extremely relevant for firms location.

3.5 Krugman and the new economic geography

Under this heading lies the contribution to regional economics initially put forward by Paul Krugman in some recent works (1991a, b, c; 1992; 1993a, b, c; 1994; 1995; 1998) and further developed by other authors such as Baldwin, Mar-

[30] This result is further generalised by Fujita and Thisse (1996) who show that the spatial location pattern of firms is the result of a "trade-off where price competition pushes firms away from each other while competition for market areas tend to pull them together" (ibid, p. 25).

[31] "Capitalisation means that the price of land reflects the public services, local taxes and transport costs incurred by the occupant" (Scotchmer and Thisse 1992, p. 281).

tin, Ottaviano, Puga, and Venables[32]. In the above mentioned papers, Krugman applies the original framework he developed for international trade theory (Krugman 1990) to two main problems of spatial economics[33]: the existence of agglomeration economies and the determinants of firms' location. He firstly explains the existence of clustering phenomena (which he calls "geographic concentration") in terms of the interactions of increasing returns, transportation costs and factor mobility.

In a first model[34], he describes an economy divided into two sectors (agriculture and industry) and two locations (East and West). There are two sector-specific inputs (farmers and workers) and only workers are free to move wherever they enjoy an higher real income; farmers are immobile. Agriculture displays constant returns to scale and no transportation costs, while industry exhibits increasing returns to scale, transportation cost for its products and Chamberlinian monopolistic competition with free entry as market structure[35].

The model is based on the relationship between three forces: two of them are centripetal (firms want to locate close to the largest possible market, and workers want to have access to the larger number of goods) and one is centrifugal (firms want to serve the peripheral agricultural market). The existence (and the persistence) of phenomena such as the "geographic concentration" of industrial activity therefore depends crucially on the interaction of three parameters describing: the share of manufactured goods in total expenditure, the level of transportation costs, and the extent of scale economies[36].

The location decision of firms is explained through the interaction of two elements: an initial historical accident and a cumulative process which is built by translating some of the concepts originally highlighted by Marshall (1921) into a dynamic analytical framework[37]. Krugman explicitly focuses on the advantages of labour market pooling and of local supply of specialised inputs to cover a wider range of industries; however he does also cover in the analysis the case of high-

[32] We mainly refer to the following works: Baldwin (1997a and 1997b), Baldwin et al. (1997), Venables (1995 and 1996), Puga and Venables (1996 and 1998), Ottaviano and Puga (1997).

[33] For further details on the relationship between international trade theory and location theory, see Krugman (1993c).

[34] Sketched in Krugman (1991a) and analytically discussed in Krugman (1991b).

[35] This model heavily relies on the monopolistic competition model of Dixit and Stiglitz (1977).

[36] If transport costs are high a/o the share of expenditure in manufactured good is low, or scale economies are weak, no clustering will take place. However when scale economies are high enough firms will cluster and, because of transport costs, they will cluster where the demand is larger.

[37] A dynamic model approach to the process of industrial agglomeration — originally outlined by Marshall (1921) in *Industry and Trade* — had been already proposed by Paul David. The Marshallian Dynamics of Industrial Localization: Chicago, 1850 - 1890 is the title of a paper he presented in 1984 and Marshallian Factors, Market Externalities and the Dynamics of Industrial Localisation is the title of a 1988 working paper by David and Rosenbloom, published in 1990.

tech firms, by considering pure technological externalities and informational spill-overs as agglomerating factors[38].

The specific issue of the relationships between innovations, high-tech industries and regional/urban economics is specifically examined in another work (Brezis and Krugman 1993). In this article, which is explicitly entitled "Technology and the Life Cycle of Cities" — but whose conclusions, we think, can easily be extended to other geographical entities such as high-tech clusters —, they explain the overtaking of leading cities by upstart metropolitan areas, which often occurs during time of major technological change, in terms of localised learning by doing.

"When a new technology for which the accumulated experience is irrelevant is introduced, older centres prefer to stay with a technology in which they are more efficient. New centres, however, turn to new technology, and are competitive despite the raw state of that technology because of their lower land rents and wages. Over time, as the new technology matures, the established cities are overtaken" (Brezis and Krugman 1993, p. 1). This paper gives theoretical support to the empirical evidence which shows similar life cycles of technologies and territories. When a major technological innovation spreads out, new high-tech clusters appear and — if the innovation is commercially successful — they develop and consolidate often at the expenses of older historical sites. The names of the area and of the products become therefore inseparable: Detroit and the standardised car in the early 1900s, Palo Alto (Silicon Valley) and the semiconductors in the mid-1950s, Boston Route 128 and the minicomputer in the early 1980s.

Other contributors to the new economic geography approach have stressed further interesting features such as the possibility of obtaining industrial clustering through capital mobility and investment dynamics, even when the labour force is immobile (Baldwin 1997b); the role played by industry spillovers, through a series of industrialisation waves, in the diffusion of manufacturing activities from one region to neighbouring ones (Puga and Venables 1996); the influence of forward and backward linkages (Venables 1996).

An undeniable merit of Krugman's work mainly lies in the recognition that the analysis of locational choice and actions of agents is part of the core of economic theory (and not just a very peculiar curiosity). Furthermore his approach has contributed to the recognition of phenomena such as multiple equilibria, path dependency, self-fulfilling expectations and historical accidents within the main stream neo-classical economics. His work seems also to suggest that very traditional and common economic reasons (related to economies of scales, labour force location

[38] However it must be remembered that Krugman is rather sceptical on the effective role played by knowledge spillovers in determining the formation of cluster. "(I) have chosen to put pure technological externalities last, not first, for several reason. First, it is an empirical fact that many of the industries which are highly localised within the United States (...) are nothing like high-technology sectors. (...) Second, as a matter of principle I think we should focus first on the kinds of external economies that can be modelled other than by assumption. (...) Knowledge flows (...) are invisible; they leave no paper trail by which they may be measured and tracked, and there is nothing to prevent the theorist from assuming anything about them that she likes" (Krugman 1991a, p. 53).

and mobility, and transport costs) may help in explaining the locational pattern of high-tech firms better than peculiar industry specific ones[39].

However it must be noted that, in his effort to put himself aside from the community of regional scientists and urban economists[40], he has not explicitly included in the analysis some crucial features that characterise spatial economics as such. The first and most relevant of these features being congestion and agglomeration diseconomies which, sometimes, can put a halt to the development of the most promising industrial clusters[41].

3.6 Industrial geography

This approach, put forward in the early Eighties by a group of scholars[42] (mainly geographers and planners based in California), aims at stressing the causal relationship which exists between the internal (both technological and economic) dynamics of a capitalistic economy and the territorial pattern of industrialisation[43].

According to this approach, industrial location patterns are created through "the process of growth rather than through a process of efficient allocation of plants across a static economic landscape" (Storper and Walker 1989, p. 70). In contrast to Weberian location theory, industries are therefore assumed capable of "generating their own conditions of growth in place, by making factors of produc-

39 "So while I am sure that true technological spillovers play an important role in the localisation of some industries, one should not assume that this is the typical reason — even in the high-technology industries themselves" (Krugman 1991a, p. 54).

40 This point is clearly expressed in two distinct quotations which can be found in a later contribution: "Instead of a deep body of theoretical work, what Isard ended up creating was an eclectic field: regional science. Regional science is not an unified subject. It is best described as *a collection of tools*, some crude, some fairly sophisticated, which can help someone who needs an answer to practical problems involving spatial issues that will not wait until we have a good theory. (...) Over the last few years I have been gradually constructing a model of a spatial economy that relies on the Dixit-Stiglitz approach to monopolistic competition to 'sterilize' the problem of imperfect competition. I do not clam that this approach is the only way to do spatial economics, or even that is a wholly satisfactory model. What I do claim is that the model demonstrates the feasibility of telling the kind of stories that are needed to do meaningful economic geography in a way that mainstream economists can live with" (Krugman 1995, pp. 57-60).

41 As explicitly and honestly recognised by Krugman: "But all not factors are mobile, and the presence of immobile factors provides the centrifugal force that works against agglomeration. In principle, one should include urban land rents as part of the story; in the models I have worked out so far, however, this force is disregarded. Instead, the only force working against agglomeration is the incentive to set up new facilities to serve a dispersed agricultural hinterland" (Krugman 1995, p. 91).

42 Bennet Harrison, Allen J. Scott, Michael Storper, and Richard Walker.

43 "The basic patterns of industry location and regional growth can be produced by processes endogenous to capitalistic industrialisation, rather than by the exogenous placement of resources and consumers" (Storper and Walker 1989, p. 70).

tion come to them or causing factors supplies to come into being where they did not exist before" (ibid., p.71). In other words, according to the geographical industrialisation approach, industries produce regions and are capable of creating their own geography (Scott 1993; Scott and Storper 1992; Salais and Storper 1992 and 1993, Storper 1997).

This line of reasoning — which explicitly contrasts with the established literature[44] and rejects precise analytic modelling — assumes that the spatial dynamic of industry growth follows a regular pattern divided into four main phases: localisation (a new fast growing industry creates its own location conditions, thanks to its factor-creating and factor-attracting power); clustering (dynamic economies of productions, both internal and external to firms, lead to spatial concentration); dispersal (as the industry grows new plants are located further away from established industry centres, both as a response to increasing congestion and as a growth strategy through "expansive periphery"); and shifting centre (periodic convulsions lead to the crisis of established sites and, in the long-run, to shifts in the centres of industrial activity).

Time plays an important part in this approach. In the early stages of a new industry, firms can settle on any of a wide variety of locations and avoid existing industrial agglomerations "because they profit from dynamic economies of production, accelerated investment flows and labour influx that are not necessarily dependent on the activities of firms in other (older) sectors" (Storper and Walker 1989, p. 75). These moments of enhanced locational freedom are called windows of locational opportunity.

It may be also emphasised that the industrial geography approach shares several features with a wider stream of theoretical and applied contributions[45] which, since they trade off analytical rigour against detailed descriptions of clustering phenomena, have been used in this thesis as background reference for the applied analysis of chapter 6 and the study of policy implication described in chapter 7.

For the purpose of the current analysis, this approach has the merit of attempting to unify industrial location and regional development in a single explanatory framework. Furthermore it focuses explicitly on the locational dynamics of innovative industries and firms and accounts for some of the main features of regional development: uneven territorial distribution, mixture of inertia and instability, relevance of local and global interdependencies.

---

[44] "We proposed to thoroughly rewrite location theory from the standpoint of political economy (...). Our primary aim is to move towards a synthesis to replace the neoclassical one locked into position by Isard (...). Neo-classical economic models need to be replaced at the foundation" (Storper and Walker 1989, p. 3)

[45] Which ranges from the French "regulation" school (Benko and Lipietz 1992; Boyer and Saillard 1995) to the Italian tradition of industrial districts analysis and local systems of productions (Bagnasco 1977, Becattini 1987; Garofoli 1991; Bramanti and Maggioni 1997), from the GREMI approach (Aydalot and Keeble 1988; Camagni 1991; Ratti et al. 1997) to the Scandinavian group for industrial network analysis (Hakansson 1987 and 1989; Maskell 1997; Maskell et al. 1998), from the new sprouts of "traditional" economic geography (Taylor and Conti 1997; Conti et al. 1995) to the political economy of cities and regions (Crouch and Streeck 1996; Keating and Loughlin 1997).

Its main limitations rest in the absence of any formal modelling and in the consequent impossibility of serious empirical testing. However if this literature is used as the theoretical framework underpinning the modelling techniques and the empirical analysis put forward by the "population ecology" approach (described in section 3.12), these drawbacks can be easily overcome.

3.7 Technological infrastructure approach

This very pragmatic approach aims to explain the spatial distribution of innovations in terms of the availability of well developed technological infrastructure, empirically defined in terms of the agglomerations of four indicators: i) networks of firms in related industries; ii) university R&D; iii) industrial R&D; iv) business-service firms (Feldman 1994). This recent approach, which stems from the applied research school of Acs and Audretsch and it is mainly associated with the works of Feldman (Acs, Audretsch and Feldman 1994; Feldman 1994; Feldman and Florida 1994), focuses on the empirical testing of the relevance of externalities and agglomerative forces in determining the location of high-tech firms measured, indirectly, through the size of R&D outlays and the number of innovations produced and/or patents awarded[46].

Although their explicit theoretical underpinnings are on the evolutionary side of the analysis of technological change (several references are to works by David, Dosi Freeman and Rosenberg), it seems more sensible to classify such work as a sort of indirect test of the orthodox models illustrated in section 3.3.1 stressing the role of generic spatial externalities (or urbanisation/regionalisation economies)[47] in causing the clustering of innovative activities.

It must be noted that the main interest of Feldman's research is the "geography of *innovation*"; therefore the location of innovative firms is considered only as a mere dependent variable able to partially explain the spatial distribution of product innovation. However the interest for the problem at study lies in the recognition of the importance of spatial aspects in the innovative process[48] and in the exposition of an exhaustive series of data, variables, proxies and empirical test to validate the hypothesis.

[46] For its mainly empirical character, this approach will be briefly reviewed in chapter 5.

[47] Although the denomination of a dependent variable (networks of firms) would suggest an attempt at measuring the direct interactions or communication sources of agglomeration, this is not actually the case. The indicator chosen to quantify this variable is in fact the mere value added of related industries in the region.

[48] "Innovation is no longer the province of the inventor, the risk-taking entrepreneur, the insightful venture capitalist or the large resource-rich corporation. Innovation instead has its sources in a broader social and economic institutions welded into a technological infrastructure for innovation. It is in this fundamental sense that geography organises the innovation process and helps sustain the spatially uneven growth and progress of advanced technological economies" (Feldman and Florida 1994).

3.8 Porter's competitive advantages

Even though the initial focus of Porter's analysis (Porter 1980 and 1985) was on firms and firm's strategy, since 1989 he has become interested in the analysis of two main environments where firms must decide and act, these being the industrial sector and the nation. In *The Competitive Advantage of Nations*, he argues that the competitive advantage of a national industry crucially depends on four main determinants: "(i) factor conditions (e.g. the nation's position in the factors of production, such as skilled labour or infrastructure, necessary to compete in a given industry; (ii) demand conditions (e.g. the nature of home demand for the industry's product or service; (iii) related and supporting industries (e.g. the presence or absence in the nation of supplier industries and related industries that are internationally competitive); (iv) firm strategy, structure and rivalry (e.g. the conditions in the nation governing how companies are created, organised, and managed, and the nature of domestic rivalry" (Porter 1990, p. 71).

When the analysis focuses on a series of industry-specific case-studies from ten different countries, the territorial level of analysis sometimes shifts from nations down to local industrial clusters[49]. In successive works, Porter adapts the "diamond" of national competitive advantages to the local framework by stating that "Regional clusters grow because of several factors: concentration of highly specialised knowledge, inputs and institutions; the motivational benefits of local competition; and often the presence of sophisticated local demand for a product or a service". (Porter 1996, p. 87).

His contribution has been summarised by Glaeser et al. (1992) as follows: local competition and industrial specialisation are the main engines of clusters growth[50]; and tested in an empirical analysis on 170 large US cities. A similar analysis, relative to high-tech industries, is performed in chapter 6 of this thesis.

3.9 Jacobs and the economic history of cities

The reasons for including Jacobs contribution in this survey of the literature are twofold. On the one hand, her name is used by Glaeser et al. (1992), to represent an alternative explanation of why cities grow in the above mentioned empirical exercise. On the other hand, her contributions (Jacobs 1964, 1969, and 1984) show that some remarkable results can be obtained by mixing, in the right proportion, an extensive use of historical cases and anecdotes and a sharp economic intuition.

One of the main ideas of Jacobs is that variety and diversity of geographically proximate industries promote growth rather than geographical specialisation. This is expressed with vividness in Jacobs (1969, p. 48): "Our remote ancestors did not

---

49 This dynamics is particularly evident in the Italian and in the German case studies in Porter (1990).

50 "Knowledge spillovers in specialized, geographically concentrated industries stimulate growth. (Porter) insists, however, that local competition, as opposed to local monopoly, fosters the pursuit and rapid adoption of innovation" (Glaeser et al. 1992, p. 1127-8).

expand their economies much by simply doing more of what they had already been doing (...). They expand their economies by adding new kind of work. So do we. Innovating economies expand and develop. Economies that do not add new kinds of goods and services, but continue only to repeat old work, do not expand much nor do they, by definition, develop". Jacobs is well aware that industrial specialisation brings short run efficiency (through localisation economies); however she contrasts such efficiency with the long run resilience which derives from a diversified industrial structure[51]. In a sense, Jacobs is assuming that urbanisation economies are better than localisation economies for ensuring a long run growth path.

Another keen intuition of Jacobs is that small average firm size (and indirectly competition) foster the process of growth and development. "Offhand one may suppose that large organisations with their many divisions of labour would be much more prolific at adding new work to old than would small organisations. But this is not so. In a large organisation nearly all the divisions of labour, no matter how many there are, must necessarily be sterile in this respect" (Jacobs 1969, p. 71).

Thus, according to Jacobs, a competitive environment with a large number of small firms — belonging to different industrial sectors — is the engine of growth and, above all, long run development. As a consequence, high-tech clusters — characterised by a differentiated industrial structure and by a low average size of establishments — must perform better than specialised clusters where large firms are predominant. Such an assumption will be empirically tested in chapter 6.

3.10 Diffusion theory

A large part of the literature dealing with the economics of technological change concerns the study of the diffusion process of innovation. However the process of diffusion is a crucial feature in other scientific fields, from medicine (the spread of a virus), to ecology (the spread of a new species), to sociology (the spread of rumours). In general the main issues at study in diffusion theory is to explain why, once a new phenomenon (technological innovation, virus, biological species, rumour) has appeared, it takes time, for such a phenomenon, to reach the entire population and/or completely fill the environment. It seemed therefore natural to the author to think about the location process of a certain type of firms (and about the birth and development of an high-tech cluster) as a diffusion process where the main question to be answered is: why, once a new location has appeared and has proven to be superior, it is not "invaded" by all potential entrants? While section 4.5 deals explicitly with this very question, this section presents a very brief sur-

[51] This point is illustrated through the example of two English manufacturing cities in the middle of the 19th century: Manchester and Birmingham. "Manchester (...) was indeed efficient and Birmingham was not. Manchester had acquired the efficiency of a company town. Birmingham had retained something different: a high rate of development work. (...) Today, only two cities in all Britain remain economically vigorous and prosperous. One is London. The second is Birmingham" (Jacobs 1969, p. 87-88).

vey of the main ideas developed in the economic literature on the diffusion of technological innovation.

The epidemic model (Grilliches 1957; Bass 1969) emphasise the role played by information spreading in the diffusion process of an innovation. If one assumes that information diffuses through contacts, and that these contacts are random, then at any moment of time the rate of diffusion of an innovation is proportional both to the fraction of actual users and to the fraction of potential users. The diffusion pattern follows a logistic curve (i.e. a particular type of S-shaped curve in which the rate of diffusion first increases until an inflection point[52] is reached and then decreases as the curve approaches the equilibrium level where the all agents are actual adopters). The main limitation of such an approach — as pointed out by Davies (1979), Stoneman (1983 and 1986), Coombs et al. (1987), Karshenas and Stoneman (1995) — lies in the fact that all adopters are assumed to be homogenous, that both the population of potential adopters and the features of the innovation are assumed to be the same throughout the entire diffusion process, and that only the demand side of the diffusion process is taken into account, while the role played by the suppliers of the innovation is overlooked. Mansfield (1968), reinterpreting the epidemic model, argues that the diffusion process is determined by the uncertainty on the performance characteristics of the innovation and hypothesises the speed of diffusion to be positively related to the profitability of the innovation and negatively related to the size of the investment required. Other authors (such as Chow 1967; Glaister 1972) have proposed various modifications to the basic framework to accommodate a number of limitations. However all these models assume the existence of an equilibrium level (when all potential adopters become actual ones), underline the information constraint in any moment of time, and describe the "diffusion path (as) a disequilibrium approach to that end point" (Karshenas and Stoneman 1995, p. 273).

Alternative interpretations of the diffusion process (equilibrium models) assume that, at any moment in time, there is perfect information in the economy on the existence and nature of new technologies. However each firm, before deciding whether or not to adopt, must compare the benefits and the costs of adoption. "As time goes on the cost of adoption (or the benefits from adoption) change and, as they do so, usage of the technology extends" (ibid, p. 274). In the "rank effects" models (David 1969; Ireland and Stoneman 1986) it is assumed that the heterogeneity of potential entrants causes different returns from adoptions and, indirectly, different dates of adoption. In the "stock effects" models (Reinganum 1981; Quirmbach 1986) the benefits from acquisition to the marginal adopters are assumed to decrease as the number of previous adopters increases. In the "order effects" models it is argued that adoption benefits to a firm depends on its position on the order of adoption (on the basis of a "first come, better served" criterion). Karshenas and Stoneman (1993) put these three approaches in one encompassing model and are able to empirically estimate the relative influence of each factor in determining the diffusion process of computer numerically controlled machine tools in the UK. A modified version of such a model, able to describe the dynamics of the location process, will be presented in section 4.5.

---

[52] Corresponding to half the population of potential adopters.

3.11 Path dependence, lock-in and informational cascades

This approach derives from the application of original lock-in models — first proposed by Arthur (1983, 1989) to describe a context where agents must choose between different technologies competing for adoption — by Arthur to the field of industrial location (Arthur 1988, 1989, 1990).

In the most simple spatial framework, the basic mechanism at work is as follows: benefits arising to firms from location can be split into two parts: geographical benefits (depending only on the specific location's characteristics) and agglomeration benefits (depending on the number of previous locations). Firms are assumed to have different geographical preferences (or tastes)[53] over a finite number of regions. The function representing agglomeration benefits is either monotonously increasing (assuming unbounded economies of agglomeration) or concave, i.e. firstly increasing then decreasing (allowing for the existence of a threshold size after which diseconomies of agglomeration prevail). The model is then completed by a randomly chosen series of outsider firms, which originates a sequence of locations[54].

When agglomeration benefits are set to zero, and firms locational preferences are independent[55], then a definite pattern of industrial settlement[56] will emerge which will not be effected by any different sequence of historical events in the formation of the industry[57].

When agglomeration economies exist and are unbounded, as the industry enlarges, one location will emerge as "spatial monopolist" and get the entire set of subsequent locational choices. Which location would be chosen cannot be predicted in advance and depends on the level of geographical benefits and on the order of entry of firm types (where a key role is played by early locators' locational

---

[53] I. e. different types of firms, in absence of agglomeration benefits, would choose different regions for their location.

[54] A forerunner of this approach may well be considered Schelling (1978) with his works concentrating on the aggregate effects of individual actions as the title of his most famous book *Micromotives and Macrobehaviours* vividly suggest.

[55] In the modelling framework proposed by Arthur the independence of location derives from the absence of agglomeration benefits; on the contrary in the "informational cascades" literature, location decisions of firms are shown to be dependent even when agglomeration benefits are absent because of their informational contents.

[56] Where the locational share of each region is equal to the proportion of firms with that given locational preference.

[57] In Arthur (1988) an alternative location dynamics is presented in which each region is initially inhabited by one firm and the industry develops through new firms spinning-off from parent firms one at a time. If this is the case, geographical benefits are set to zero and "firms are added incrementally to the region with probabilities exactly equal to the proportion of firms in each region at that time" (Arthur 1988, p. 88). This locational process will be completely dominated by chance, or historical events, and any consistent locational pattern (i.e. where all regional shares sum to 1) is likely as any other.

tastes). "Attractiveness, interacting with historical accidents of choice-order, determines the outcome" (Arthur 1990, p. 243).

Much more interesting is the case where both agglomeration economies and diseconomies are modelled. In this case several results are possible depending on the values of key parameters describing the heterogeneity of firms' locational tastes, the sequence of locations, and the force of agglomeration economies. In particular it may happen that, when locational tastes are very relevant and/or location heterogeneity is infinite, "all locations (will) share the industry, with probability one, in the proportions that would have occurred if agglomeration economies were absent" (ibid. p. 246). Alternatively, when locational preference are tightly clustered (i.e. there is almost no locational heterogeneity) and/or agglomeration effects at the outset are strong, then "one location takes all the industry with probability one" (ibid. p. 246).

A further interesting (and somehow counterintuitive) result deals with agglomeration economies being responsible for spatial separation and firms spreading over a territory. This fact can be easily explained when it is assumed that contiguous locations grant similar geographical benefits to resident firms. If this is the case, and we allow for agglomeration economies, then locations with large numbers of firms will be preferred to geographically similar neighbouring locations with fewer firms. "Locations with large numbers of firms therefore cast an 'agglomeration shadow' in which little or no settlement takes place. This cause the separation of the industry" (ibid. p. 247).

These results are further re-enforced by adding those deriving from a distinct, but converging, stream of literature dealing with "informational cascades" and "herd behaviour" (Banerjee 1992; Bikhchandani et al. 1992; Hirshleifer 1993; Hirshleifer and Welch 1994). In these papers, lock-in processes arise even in absence of pay-off interactions (i.e. when one agent's action does not directly increase the benefits to other agents doing the same thing) because of limited private information and of a tendency of extracting further information from other agents actions. In such a situation, it may well be optimal for an agent, having observed the actions of those ahead of him, to follow the behaviour of the preceding agent without regard to his own information. This causes a multiplicity of outcomes, volatility of the equilibrium pattern of choices across several plays of the same game, and the possibility for the system to be locked-in to inefficient equilibria.

Although these latter works do not make explicit reference to locational issues, we think that this approach may be very relevant for the topic at study. According to this modelling framework, the number of previous locations in a given area is seen by outsider firms as a signal of profitability[58], and thus influences the subsequent location choices even when agglomeration economies are absent or, even more convincingly, when they are unknown to outsiders. For this reason, some insights arising from this section have been incorporated in the ecological models used in chapter 4.

[58] The number of located firms is indeed the more evident and less costly signal of the profitability of a given location; however firms do also take into account that the ratio signal to noise, for such an issue, can sometimes be very small.

3.12 Biological and ecological models

Under this heading we look at two different streams of literature which, in various ways, use biological and ecological analytical frameworks to analyse the growth process of social and economic collective agents (these being: cities, regions, industries, firms, unions). In particular we will briefly review the dynamic models of self-organising urban growth (Allen and Sanglier 1978, 1981a and 1981b; Allen 1980), and the mathematical ecology of cities (Dendrinos and Mullally 1985; Dendrinos and Sonis 1986; Sonis 1986; Dendrinos and Rosser 1992; Dendrinos 1996; Reggiani 1997), regions (Maggioni 1993; Nijkamp and Reggiani 1992 and 1998; Folloni and Maggioni 1994; Staber 1997; Gambarotto and Maggioni 1998), industries (Brittain and Wholey 1988; Maggioni 1993; Staber 1997), technologies (Andersen 1994) and organisations (Carroll 1988; Hannan and Freeman 1989; Hannan and Carroll 1992).

Dynamic models of self-organising urban systems developed at the end of the 70s within the multi-disciplinary school of Prigogine. In particular, Allen and Sanglier were the first to model the birth and the development of an integrated (and hierarchically ordered) system of cities according to the self-organisation paradigm. The main assumptions of this paradigm are: i) any system is subject to external flows of energy or matter that more than compensate the internal entropy production; ii) relationships between the elements of the system are non linear.

These models of urban dynamics (Allen 1976 and 1980; Allen and Sanglier 1978, 1981a and 1981b) are based on two variables (number of residents and number of jobs available in each location), two types of agents (employers and employees), a series of different economic activities (or functions as in the original model of Christaller, 1933) that are initially introduced at random in different cities, a basic mechanism (workers migrate searching for employment, employers set labour demand according to the size of the market they serve), and two non linear interactions (first, the "urban multiplier", linking positively the installation of an exporting activity, the level of local population and local potential employment capacity; second, the external economies cycle, which links in a loop: concentration of employment, provision of common infrastructures, further concentration of economic activities).

The results, obtained by simulating the model with different values for key parameters, recall those of Christaller: the system gradually evolves from an undifferentiated set of equal cities to an hierarchic distribution of cities ordered by population size and economic function provided. However the spatial symmetry is lost and several different outcomes may arise depending on historical accidents and on parameters describing the non linear relationships. These simulations-which represents a fruitful interaction between classical location theory and the theory of non linear dynamics — enlarge the set of possible equilibria of established locational models (thus reducing the generality of their results) and highlight the possible emergence of chaotic dynamics.

The stream of literature devoted to the analysis of ecological models gradually evolved, at the beginning of the Eighties, from a conviction that urban and regional problems have an intrinsically dynamic, complex, and relational nature. For this reason it seemed possible to draw a methodological analogy between the development of a spatial systems (a city or a region), dependent on the external environment (national and international economic conditions) and on the behaviours of other similar entities (i.e. other cities) operating in the same environment, and the development of a species of animals, dependent on the conditions of the natural environment and on the behaviour of other species. This approach can be considered as complementary to the paradigm of evolutionary economics (Freeman,1982; Dosi et al. 1988; Metcalfe 1994). While evolutionary models underline the role of variety-creation mechanism (e.g. the innovation), ecological models stress the role played by the selective mechanism: the environment (which in turn is made of external conditions and inter and intra specific interactions). Usually these models initially describe the laws of development of a city, region or industry in isolation (intra-specific ecology); then they analyse the possible forms of interactions exhibited by systems of cities, regions or industries (inter-specific ecology); and finally they deal with more general questions about the relationships between complexity and stability of urban, regional or industrial systems. A very similar approach will be used in chapter 4, when describing the evolution of a high-tech firm cluster, firstly by modelling its genesis and development in isolation, then allowing for the interactions with other clusters, finally introducing the influence of external exogenous macroeconomic shocks.

The mathematical framework of these models is rather simple, based on differential and difference equations — and has been sketched in its core elements by Lotka and Volterra in the late 1920s, early 1930s — but due to the features of dynamics and complexity it shows a worrying tendency towards analytical difficulty and impossibility, thus the need for numerical simulations. The number of different models that can be derived from this very basic theoretical skeleton is amazingly large and include the possibility of describing and analysing competitive as well as co-operative interactions.

3.13 Concluding remarks

The survey of the theoretical literature, presented in this chapter, has shown that different approaches highlight different aspects of the issue of firms' location and agglomeration. This section and the following table aims at summarising the chapter's main findings by describing, for each theoretical approach, the main advantages and drawbacks in explaining high-tech clusters birth, development and evolution.

Table 3.1 Summary of theoretical approaches

Theoretical approaches	*Advantages*	*Drawbacks*	*How clustering is explained*
Least cost and land utilisation	supply side orientation; distance related variables; multiple equilibria	overlooking of demand side; perfect competition	resources location; labour force pool; agglomeration economies
Demand side and game theory	role of strategic behaviour; multiple equilibria; explicitation of hypotheses	non co-operative firms' interactions; technological synergies	comparative shopping; maximisation of market power
General location equilibrium	non price interactions; monopolistic competition	lack of a unifying framework	demand-supply interactions among firms and between firms and households
New economic geography	monopolistic competition; relation between the life cycles of technologies and cities.	inexistence of congestion dynamics	labour market pooling; local supply of inputs; preference for variety
Industrial geography	existence of windows of locational opportunity; industries produce regions	no explicit formal modelling	dynamic economies of production; horizontal integration
Technological infrastructure	serious empirical exercises	weak relation to a specific theory	localisation economies; industrial and university R&D; business service firms
Porter's competitive advantages	use of case- studies; heuristic and pragmatic approach	must be reduced in order to be empirically tested	localisation economies; beneficial effects of local competition; local concentration of demand
Jacobs and the economic history of cities	mix of historical cases and economic intuition	must be reduced in order to be empirically tested	urbanisation economies and local competition
Diffusion theory	explains why location is a lengthy process. Allows for both equilibrium and disequilibrium analyses	equilibrium analyses needs firm level data for testing	information diffusion; network externalities
Lock-in and informational cascades	existence of path-dependency, lock-in and sub-optimality of equilibria	results are mainly obtained through simulation.	informational cascades; agglomeration economies; locational preference
Biology and population ecology	self organisation; non linear and complex dynamics; multiple equilibria	absence of strategic behaviour; need for simulations	inter-specific synergetic relationships

Within the classical location theory, the "least cost" approach underlines the supply side factors in determining location and clustering. Its distinctive emphasis on transport costs[59], labour location, and agglomeration economies provides the guideposts for our analysis. The main drawbacks — being the lack of analysis of the demand side and of producers' strategic behaviours — are partially compensated for by the contribution belonging to the "market area" and "locational interdependence" approaches which show that clustering is possible even in absence of agglomeration economies through strategic locational behaviour aimed at maximising market power[60]. The "land utilisation" approach underlines the role played by rent in determining real life industrial locations.

The general equilibrium approach, especially in its sub-divisions dealing with non price interactions and monopolistic competition raises some interesting points. Firstly: agglomeration occurs because firms profit from face-to face interaction, clustering thus becomes a valid alternative to vertical and horizontal integration. Secondly, the observation of the emergence of industrial clustering (e.g. spatial concentration of similar firm) reduces the uncertainty which is inherent to any locational choice. Thirdly, agglomeration economies are either produced by concentration of specialised business service firms, derive from market exchange of intermediate goods, or may arise from the interactions of firms and households.

Game theory re-visited location theory, limiting the generality of results and adding new insights on the strategic side of price and non price spatial competition. Some interesting results for the issue at study could be obtained by focusing on co-operative games (stressing the role played by coalitions) or by considering the interactions between different local authorities trying to sustain the industrial development of their area each one at the expense of the others.

Krugman's new economic geography stresses the role of labour force pools and localised sources of specialised suppliers (in conjunction with increasing returns to scale) in determining agglomeration economies and consequently clustering. More importantly this approach also sheds light on the relationship between radical innovation and birth of new urban centres and industrial clusters.

The Californian school of industrial geography polemically states that location factors are unimportant. Industries produce regions and not *vice versa*, by attracting relevant location factors which, in the medium run, are all mobile. Locational patterns and industrial clusters are the result of historical accidents and of the inner spatial dynamics of capitalist economies, which can be divided into four phases: localisation, clustering, dispersal and shifting centres.

The technological infrastructure approach offers an empirical confirmation of the relevance of geographical clustering and agglomeration economies for innovative activity. Product innovation is more likely where there are developed tech-

---

[59] Which, as suggested in section 3.2.1, can be re-interpreted as distance-related increasing difficulties in the transfer of specific, complex, cumulative technology (or, more generally, of any tacit, and non codified knowledge).

[60] The most famous result being the "principle of minimum differentiation" illustrated by Hotelling (1929).

nological infrastructures (measured by number of related firms, size of university and industrial R&D and number of business service firms). If one assume that the best location for an high-tech firm is a site where the innovation rate is above average, then the endowment of technological infrastructures becomes the most important location factor for innovative firms.

The lock-in approach is extremely powerful in its simple analytical framework: locational preference, plus increasing returns to location (limited or unlimited), plus a random order of location choice gives an incredible variety of possible results, from even distributions to spatial monopoly. A further interesting result (complementing Hotelling's principle of minimum differentiation) is that agglomeration economies can, in special cases, cause a particular form of geographical dispersion (locational shadowing).

The contributions by Porter and Jacobs put forward two alternative explanations for the process generating the growth of an industrial cluster and a city. The first stresses the role played by intra-industry localisation economies and local competition, while the second underlines the influence of industrial variety and urbanisation economies, in conjunction with small average firms size.

The bio-ecological approach highlights the issues of spatial self-organisation, non linearities, plurality of equilibria and of level of analysis, non competitive relationships and has called for numerical simulations to understand complex dynamic systems.

Being interested in the analysis of high-tech clusters, we have selected the most relevant suggestions within each approach to delineate an original research scheme analysing clustering from a twofold perspective. The first is a macro perspective, the second is a micro one. The former focuses on the overall dynamics of high-tech cluster: why they occur, where they develop and how they evolve. The latter concentrates on the different factors which determine the location decisions of a firm. In the following section we discuss these two perspectives which will be built upon in the next chapter.

3.14 Moving forward

One of the most relevant features of industrial development, as shown in chapter 2, lies in its unequal spatial distribution. Other clear empirical evidence shows that different industries choose different geographical locations and sites in which to cluster. Thus, one would assume that firms location processes and industrial cluster dynamics should constitute the core of any analysis of locational issues.

On the contrary, as it has been shown in the pages above, these two intertwined phenomena have been analysed under different headings (location theory and regional economics) which share few (if any) elements[61]. In this way, an encom-

[61] See, for example, the structure of the first volume, devoted to regional economics, of the *Handbook of Regional and Urban Economics* (Nijkamp 1986) which is divided into two separate parts: Part 1, "Location Analysis" and Part 2, "Regional Economic Models and Methods".

passing perspective which focuses on the relationship existing between firms' location decision and industrial clusters development process has been unjustly sacrificed and neglected[62].

In order to build such an encompassing framework, we need to take into consideration the existence of a methodological trade-off which states the impossibility for a model to simultaneously maximise generality, realism and precision when describing, interpreting and explaining reality. We therefore decided to use two different modelling frameworks, in order to look at the same issue from two complementary perspectives, and to try to enrich each approach with some suggestions and hints deriving from the other one.

For this reason, in sections 4.2, 4.3, and 4.4 we use an analytical framework, derived from the population ecology approach, which trades off precision against greater generality and realism to describe different possible dynamic interactions which may arise between the development path of different high-tech clusters. In this model the original ecological approach has been modified in order to introduce some micro foundations for the macro-economic dynamics. Is this way, we have been able to analyse the effects of firms profitability expectations on the development dynamics of industrial clusters. The results obtained suggest that the development or, alternatively, the underdevelopment of an area is jointly determined by the composition of potential entrants' population, by the behaviour of neighbouring clusters, by the local economic conditions and, as stressed in section 4.4, by the depth and frequency of external macro-economic shocks.

In section 4.5 we present a model, derived from diffusion theory, which trades off realism against precision and generality in order to analyse the specific decision making mechanisms involved in the location process of a profit maximising firm and to derive its effects on the overall location pattern. This model, derived from a contribution (Karshenas and Stoneman 1993) originally designed for the analysis of firms' technological adoption process, is able to describe the locational choices of high-tech firms and their effect on the development pattern of an high-tech cluster. This model is able to take into account several determinants of firms' location: previous locations (epidemic effect), individual firm's characteristics (rank effect), price and non price interactions between firms (stock and order effects). The epidemic effect takes into account the spread of information on the profitability of a specific location site; the rank effect measures the relevance of individual firm's features such as size, financial structure, management style; the stock effect describes the dynamics of agglomeration economies and diseconomies; finally, the order effect investigates the strategic components of firms' location process.

[62] With the notable exceptions of North (1955) and, more recently, of Swann (1993 and 1998).

Chapter 4 - Modelling firms' location and cluster development

Ecology is the study of patterns in nature, of how these patterns came to be, how they change in space and time, why some are more fragile than others. Population ecology is concerned with how populations interact with the environment and how these interactions give rise to the larger patterns of communities and ecosystems. The environment is more than just sun, air, earth and water: it includes other organisms which may help or hinder the survival of a species. Population ecology is also the study of how these organisms interact (...) in competition and in co-operation.

S.H. Kingsland (1985), *Modelling Nature. Episodes in the History of Population Ecology.*

4.1 Introduction

The second chapter of this thesis has shown the empirical relevance of high-tech clusters in four advanced countries, the third has presented a survey of different streams of literature which have directly dealt with, or that can be profitably used to analyse, the agglomeration dynamic and the emergence of spatial concentration of industries. This chapter is devoted to the elaboration of a new theoretical modelling framework able to explain, on the one hand, the location process of firms and, on the other, the development process of industrial clusters.

Since the issue to be investigated is twofold, we decided to use a twofold theoretical approach. For this reason this chapter is divided into two main parts. Sections 4.2, 4.3 and 4.4 study the development path of a high-tech cluster interacting with other clusters and industries, and build some micro-economic foundations of firms' behaviour. Section 4.5 looks at the location decisions from the individual firm's perspective and obtains a development path of the cluster, considered in isolation, as a consequence.

4.2 Firm location processes and the development of clusters: a macro-economic (ecologically-derived) approach

In this section we theoretically deal with the macro-economic aspects of the location process, concentrating on the patterns (and causes) of regional industrial clustering and development, on the relationship between the development of different industries in the same clusters and on the intertwined development of clusters and industries.

If one starts the analysis from this macro-economic perspective, the first issues to be addressed concern the existence and the extent of spatial clustering processes, the main questions to be answered being the following: why do firms cluster? and why do clusters have a finite size?

4.2.1 The effect of previous locations on clustering dynamics

In the population ecology literature, a generic modelling framework describes the growth process of a species in an environment as driven by its reproductive capacity and limited by the available amount of resources and the presence of other interacting species. From a mathematical viewpoint, the core of this situation can be represented by a differential equation (or by a system of differential equations) which describe the changes over time of a variable (birth minus deaths) as function of: the level of the same variable (the size of the population) at each moment in time; a ceiling level (which takes into account the limit imposed by the available amount of resources); and the level of other variables (which represent the interacting species).

Similarly, in the following sections, firms' decisions to locate into a cluster (and consequently the entry rate and the development path of the industrial cluster) are explained in terms (as a function) of the "economic mass" of the cluster, measured through the number of firms already located there.

However, these models do not merely postulate the existence of a stock-flow mechanism between the number of incumbent firms in the cluster and the entry rate, even if one may reasonably assume the number of spin-off firms, which are generated in each period within the cluster, to be a proportion of the number of firms already established there. Neither do they assume that the number of firms is the only relevant variable considered by potential entrants when taking locations decisions. Instead, as it will be shown in greater detail in section 5.1.3, the number of already located firms reduces potential entrants' search costs, signals the existence of geographic and agglomeration benefits, diffuses information, reduces uncertainty, and causes informational cascades.

What follows is thus a series of models aimed at explaining location decisions and the development path of a high-tech cluster when firms interactions (both effective and perceived) are relevant. For the sake of exposition, we first present a model describing the effects of the location decisions of monoplant firms (i.e. where firm and establishment coincide), belonging to a given industry, within an isolated cluster (i.e. when there is only one potential site for location). In the following sections, the scope of the analysis is enlarged to encompass a number of interactions between two (or more) clusters and/or industries.

4.2.2 Locational benefits and costs and the development of an industrial cluster

Firms decide to settle in a region on the basis of the expected profitability of being located there. This profitability depends on net locational benefits — obtained as the difference between gross locational benefits and costs — which, in turn, are based on both observable and unobservable elements.

For simplicity it can be assumed that, in an uncertain world — with limited information regarding local costs and revenues available to the outsiders — profitability expectations for any particular location will be based solely on the number of firms already located there (the number of previous locations being the only observable variable). Furthermore section 4.3.1 will present several other reasons for explaining why outsiders can use the cluster industrial mass (i.e. the number of located firms) as the best available indicator of a cluster's profitability and, consequently, as the main variable to be considered when taking locational decisions.

Let us assume, as in Arthur (1988 and 1990), that locational gross benefits B_{fq} for firm f locating in cluster q are composed of geographical and agglomeration benefits[1].

Geographical benefits G_{fq} depend on the intrinsic features of the site (such as the quality of local factors of production: capital k_q and labour l_q; the efficiency of the local network of specialised suppliers and business service firms s_q; and the quality of urban and industrial infrastructures u_q). Agglomeration benefits $A_{fq}(n_q)$ are a concave non monotonic function of the number of incumbents (i.e. firms already established in cluster q) n_q. Thus:

$$B_{fq} = G_{fq}\left(k_q, l_q, s_q, u_q\right) + A_{fq}(n_q) \tag{4.1}$$

The assumption of concavity and non monotonicity in A_q implies that, as the number of firms located in cluster q increases, gross benefits firstly increase be-

[1] For analytical convenience we split locational benefits in two classes: geographical and agglomeration benefits. The first class refers to those components which are unaffected by the number of incumbents; while the second refers to those components which depend on the number of incumbents. By adopting this formulation, however, we do not intend to state that agglomeration benefits refer only to spillovers of scientific and technological knowledge and know-how. On the contrary we are convinced that relevant agglomeration benefits derive also from external economies of scale in the use of local resources. The same variable (i.e. labour productivity) has a fixed geographical component, which depends on the quality of local workers, and a variable agglomerative component which depends on the number of firms already located in the region.

cause of agglomeration economies (due to productive specialisation; scientific, technical and commercial spillovers; reduction in both transport and transaction costs, increases in the quality of the local pool of skilled labour force and in the efficiency of the local credit market); then decrease when congestion more than compensates for agglomeration economies.

Locational costs c_{fq}, symmetrically, include two components: geographical costs g_{fq} (reflecting the cost structure of the cluster in terms of locally prevailing wage w_q and interest rate r_q; average price of business services d_q; and level of land rent and taxation t_q), and agglomeration costs a_q, which are assumed to be a convex non monotonic function of the number of regional incumbents n_q.

$$c_{fq} = g_{fq}\left(w_q, r_q, d_q, t_q\right) + a_{fq}(n_q) \tag{4.2}$$

The assumption of convexity and non monotonicity in a_q implies that, as the number of firms in cluster q increases, locational costs initially decrease until some "optimal" number of users for a given set of urban, industrial and environmental infrastructures and resources is reached. Then they increase due to the competition, between a larger number of firms, for a limited pool of local inputs (i.e. capital, labour, business services, land and public infrastructures) which raises their prices[2].

Net locational benefits can now be calculated as the difference between equations (4.1) and (4.2).

$$N_{fq} = B_{fq} - c_{fq} = H_{fq}\left(w_q, r_q, d_q, t_q, k_q, l_q, s_q, u_q\right) + h_{fq}\left(n_q\right) \tag{4.3a}$$

Assuming that the geographical benefits and costs do not change overtime, if we focus the analysis of the location process on the dynamics of the interactions between the level of available locational benefits, what becomes relevant for describing firms' location decisions is just the net benefit function N_{fq} in the incumbents' space. We can therefore summarise the geographic components H_{fq} with a parameter α_q, which vertically shifts the locational net benefits function, and write the following expression:

$$N_{fq} = B_{fq} - c_{fq} = \alpha_q + h_{fq}\left(n_q\right) \tag{4.3b}$$

[2] An alternative explanation for the convexity of the locational costs function for firm *f* runs as follows: the locational costs function is compose by a "fixed" and a "variable" component. The fixed part of the costs (geographic costs) decreases as the number of entrants increase; while the variable part increases (because of competition) as the number of entrants increase. The combination of these two effects produce an U-shaped (convex) cost curve as the interaction between fixed and variable costs of production in standard microeconomics textbooks. A symmetric reasoning may also explain the inverted U-shaped benefits function. This interpretation is surely more realistic than the one used in the thesis, however it is not as theoretically efficient as the other one since both components become dependent on the number of incumbents.

It easy to see that the locational net benefits function (4.3b) is always concave, since N_{fq} is equal to the difference between a concave function $B_{fq}(n_q)$ and a convex one $c_{fq}(n_q)$. In other words, each marginal firm, which enters the cluster, increases the average profitability of locating in the region only up to a threshold. After that point, any new entrant lowers the average net benefits available to each resident firm and new entrant[3].

This formulation recalls some general results, obtained in the industrial location and urban/regional economics literature (Weber 1929; Isard 1956; Richardson 1978; Papageorgiou 1979; Tauchen and Witte 1983; Miyao and Kanemoto 1987), which show the existence of an optimal dimension of a given spatial agglomeration of firms and/or households because of the concavity of the various benefits functions.

We can therefore state that — if the number of potential entrants is sufficiently large and there are no relevant entry barriers — as net locational benefits initially increase, the number of incumbent firms increases; then they decrease and are finally driven to zero.

If we assume that the number of potential entrants is not constant but it is changing overtime, than the entry pattern will be determined by the rate of birth of potential entrants, since a proportion of such firms will locate in the cluster. If we allow the number of actual entrants to be proportional to the average locational benefits available in the cluster — as a first approximation (or, alternatively, as a macroeconomic hypothesis without explicit micro-foundations[4]) — and we assume that the entry rate of firms into the cluster is proportional to the current level of locational net benefits[5], then the cluster growth will initially be fast, then will slow down and finally will stop[6].

---

[3] However, as it is made graphically evident in figure 4.2, because of the inverse U shape of the marginal benefits function, there is a range, within the number of incumbents, where marginal net benefits are already decreasing, but still higher than average ones, and average net benefits are still increasing.

[4] Another part of this story refers to the role played by spin-off firms. If we believe that average net locational benefits give incumbent firms some extra profits, workers (or, better, managers) of incumbent firms, which see these extra profits, will decide to start a new firm to appropriate them. Therefore the number of spin-off firms, which are generated by parents firms, is positively related to the level of locational benefits in the cluster, generated by the incumbent firms.

[5] It is evident that, for the moment, we are implicitly assuming firms to behave in a non-strategic manner (i.e. although their entries modify radically the locational benefits available in the region, they do not take into account such a modification) and to be totally myopic (i.e. their entries are only dependent on the current level of locational benefits and not on some discounted value of the flow of locational benefits they enjoy from the moment of entry onward). In section 4.3, we will relax some of these hypotheses.

[6] This is not to say that, from a certain moment onward, entries will not take place. On the contrary, once the equilibrium level of the region is reached, new entries are still possible, but these happen at the expenses of some incumbents which are driven out of the cluster (or, more probably, out of business). This continuous series of entries and exits

If this is the case[7], one would expect the industrial growth of the cluster to follow an S-shaped path with a slow start (when locational benefits are still low), an "explosive" central period of rapid increase (when the average net benefits in the region are highest) and a final part when the region gradually reaches and then settles to its "equilibrium"[8] size as in figure 4.1.

Figure 4.1 The development path of an industrial cluster

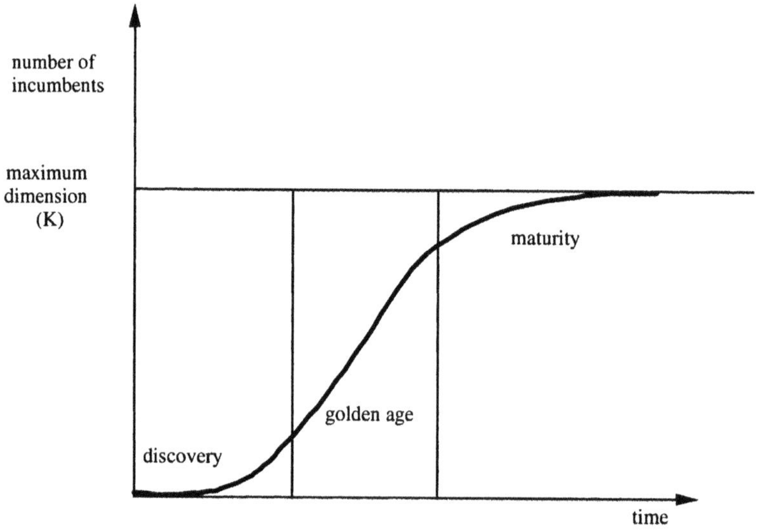

What is needed now is a formal model of the location process and the development of clusters, which will allow a rigorous testing of all these assumptions.

4.2.3 A logistic model

The simplest growth model for an industrial cluster q — which stresses the relevance of firms spatial interactions — can be expressed in the following format: "the rate of growth of the industrial mass equals the product of the individual

may determine a continuous oscillation of the cluster's industrial mass around the equilibrium level.

7 And we collected some empirical evidence supporting this story (see Maggioni 1993; Maggioni and Porro 1994; Gambarotto and Maggioni 1998).

8 It should be bore in mind, however, that this notion of equilibrium value can be changed in the long run by radical innovations, exogenous demand shocks and appropriate regional policies (see section 4.3 and chapter 7).

firm's contribution[9] to the regional population's growth and the number of firms already in the region" (Maggioni 1993).

If only agglomeration economies and positive spillovers are taken into account (and these are assumed to be constant), then each individual firm's contribution to the level of average locational benefits and, consequently, to the growth of the cluster would be equal to a constant r_q. In this case cluster industrial growth would follow an "explosive" exponential path[10], formally:

$$\frac{dn_q}{dt} = r_q\, n_q(t) \tag{4.4a}$$

Alternatively one can solve the equation for $n_q(t)$ as a function of the exogenous initial industrial mass of the cluster $n_q(0)$:

$$n_q(t) = e^{r_q t}\, n_q(0) \tag{4.4b}$$

On the other hand, if congestion and competition effects are included, then some modifications to this simple model are required to allow for some "density dependent" factors to progressively depress the level of locational benefits and to slow down the process of industrial growth of the cluster. A simple dynamic model, which takes into account these features is the logistic equation[11], which can be written as (4.5a)

$$\frac{dn_q}{dt} = r_q\, n_q(t)\left(1 - \frac{n_q(t)}{K_q}\right) \tag{4.5a}$$

where r_q is the incipient (or maximum) rate of increase and $K_q = \lim_{t \to \infty} n_q(t)$, is called the cluster "equilibrium" level[12]. Integrating equation (4.5a) and solving for $n_q(t)$ one obtains (4.5b):

$$n_q(t) = \frac{K_q n_q(0) e^{r_q t}}{K_q + n_q(0)\left(e^{r_q t} - 1\right)} \tag{4.5b}$$

---

[9] In terms of changes in the average locational net benefits, due to the interactions of agglomeration economies and diseconomies.

[10] The higher is r_q, the faster is the growth process.

[11] The logistic equation — firstly developed by Verhulst (1845) and Pearl and Reed (1920) for demographic studies, then adopted by the ecological literature since Lotka (1925) — "is the simplest model containing negative density dependence interaction. Further, it is the first two terms in a power series expansion of a more general growth model where the growth is a function of the actual size of the population" (Dendrinos and Mullally 1985, p. 38).

[12] K in the original ecological jargon is called *carrying capacity*: "a measure of the amount of renewable resources in the environment in units of the number of organisms these resources can support" (Roughgarden 1979, p. 305).

Here the individual firm's contribution to regional growth decreases as a linear function of regional population size and is equal to $r_q - \frac{r_q}{K_q} n_q(t)$ (Roughgarden 1979).

Plotting $n_q(t)$ against time yields an S-shaped curve due to the counteracting roles played by r_q and K_q. When the region is small ($n_q(t)$ is near to zero) the term in brackets in equation (4.5a) is close to one (hence the logistic equation approximately describes an exponential growth path); but as $n_q(t)$ approaches K_q, the term in brackets tends to zero, driving the growth rate to zero and terminating the entry process. Both K_q and r_q play a major role in shaping a logistic growth path: the greater is r_q the steeper is the S shaped curve, the larger K_q the higher the ceiling level of the function (and the equilibrium size of the cluster).

r_q is the incipient rate of growth. In the ecological literature it is often approximated by the difference between the birth and mortality rates of a population. This observation can be translated into the economic framework when net entry (and consequently the intrinsic rate of industrial growth of a cluster) is calculated as the difference between total entries (or start-ups) and exits (or bankruptcies) in the period considered. The same value of r_q can therefore correspond to two very different situations: a steady growing cluster where few new firms enter and no one exits, and a perturbed cluster where a high "birth" rate is almost compensated for by a high "death" rate. Hence r_q is a composite index that describes the cluster growth "potential" and the probability that firms, once entered, survive in the cluster.

K_q defines the regional industrial carrying capacity: the maximum number of profitable firms the region can sustain in isolation (i.e. when inter-regional interaction are not considered). K_q will depend upon:

1. the finite quantity of geographical benefits (which is related to the limited availability of local "resources" such as: labour, capital, land, and infrastructures);
2. the decreasing part of the agglomeration benefits function (which depends on the strategic interactions between firms: competition, congestion and lobbying of incumbents).

K_q is therefore determined by the relationship between the amount of resources (inputs) available in the region and the (technical and organisational) efficiency of incumbents in the use of these resources. Therefore in the long run K_q may change as result of the inflow of additional skilled workers, the provision of new advanced public infrastructure, the diffusion of (technical, organisational, etc.) innovations.

For a given cluster q and a given population of M_q outsider firms[13], therefore, we assume that there is an equilibrium level $K_q \leq M_q$ acting as an upper limit to the cluster's growth. In each period t, the number of entries therefore depends both on the actual number of potential entrants $K_q - n_q(t)$ (i.e. the number of outsider firms which can enter the region in time t and still make profits) and on the number of firms already located there $n_q(t)$. K_q and $n_q(t)$ in fact determine the level of average locational net benefits available to incumbent firms in each period of time.

The logistic equation thus establishes a parametric relationship between the level of locational benefits and the entry rate through the introduction of the term K_q. But what exactly is K_q and how is it determined? Is it an endogenous or exogenous parameter of the model[14]? For it to be endogenous, some assumptions about the locational cost structure of a region are needed.

In section 4.2.2 we described the functional forms of locational benefits and costs. In this section, with the help of figure 4.2 (which shows both marginal and average locational costs and locational benefits schedules), we want firstly to highlight the existence of several "optimal" sizes[15] of the region, and secondly to show how K_q is endogenously determined by the structure of locational benefits function.

Let us consider the costs and benefits derived from entering a region; for the sake of simplicity let firms outside the region experience zero locational benefits[16] and assume that geographical benefits are set to zero. Firms are assumed, for the moment, to be locationally identical (i.e. the agglomeration economies and diseconomies, locational benefits and costs are the same for every firm). Therefore we can study the behaviour of a representative firm f and analyse its average net benefits function[17].

---

[13] Composed of two main categories: firms already established outside the region and potential entrepreneurs inside the region looking for the right moment to start their own business.

[14] In the original ecological literature (Lotka 1925; May 1974; Roughgarden 1979) K is dependent on the physiological features of the population.

[15] Throughout the chapter the "industrial size" or "economic mass" of a region is approximated by the number of located firms. This index can easily be substituted by a more realistic proxy of firms' dimension (such as employment or sales). However the number of firms has an obvious advantage in its simplicity and is the best indicator when the inter-firms relationships at study (i.e. knowledge spillovers) are independent of firm's size.

[16] However, this assumption can easily be relaxed by assigning the locational benefits of the cluster a value equal to the difference between the locational benefits available outside and inside the cluster.

[17] By considering average functions we indirectly assume that some market mechanisms is at work in the cluster and makes both benefits and costs equal for each incumbent.

Figure 4.2 Agglomeration costs and benefits for incumbents and critical sizes of a cluster

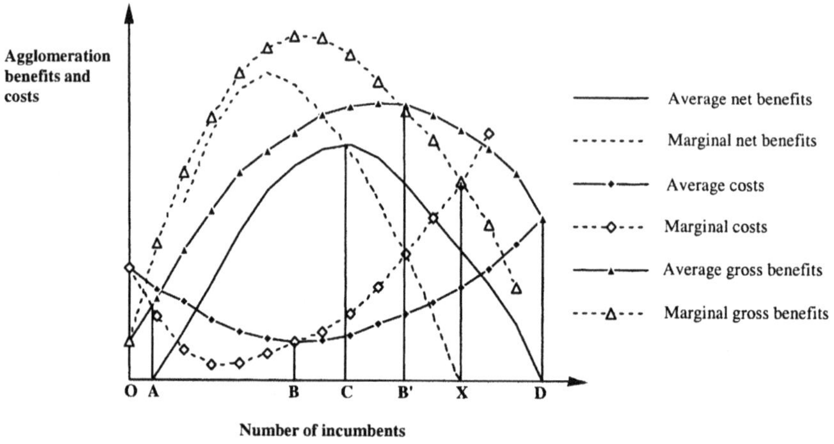

A is the minimal sustainable dimension of the cluster (i.e. where agglomeration net benefits start to be positive and, consequently, $\frac{dn_q(t)}{dt} > 0$). Prior to **A** no firm will spontaneously enter the region (because agglomeration benefits are negative). **A** can be called the "critical mass" of the region. **A** can be reached only by a group of co-ordinated firms entering together, or by direct intervention of a public authority aimed at subsidising entries until $n(t) = \mathbf{A}$.

B is the dimension where average agglomeration costs are minimum. **B'** is the cluster dimension which maximises gross average agglomeration benefits. **B** and **B'** underline the importance of analysing both costs and benefits of location to avoid harmful misrepresentation of the economic reality, as in some early contributions of location theory. Obviously, it could also be the case that **B'** < **B**.

C gives the maximum per firm net benefits (i.e. average net benefits). Up to **C** every new entrant increases (by its very entry) the average benefits of all incumbents; after **C** the average benefits decrease. **C** is therefore the optimal size of cluster for incumbent firms; however, it is neither the social efficient outcome (given that marginal benefits are still greater than marginal costs) nor the maximum possible dimension (average benefits are still positive). At **C**, several firms outside the region might still want to enter, while firms already in the region would like to deter further entries. Here we have a contrast between incumbents, outsiders and public authorities, each of them with a different view of what is the optimal outcome[18].

[18] For more on these policy issues, see chapter 7 of the book.

X is the economically efficient (i.e. social optimum) dimension. At **X** marginal costs equal marginal benefits, therefore the total benefits (number of firms times *per capita* benefits) are maximised. However, as the average benefits at **X** are still positive, some outsiders would still like to enter. Such entries would reduce the total amount of benefits available to incumbents.

D is the maximum dimension of the region (in terms of economic mass) since $B_{fq} = c_{fq}$. From **D** onwards no more net entry is deserved because, after this point, average benefits are negative and therefore there are no incentives to enter. However, new entries are still possible but these would be at the expense of some incumbents who would be driven out of the cluster[19]. Thus **D** identifies the ceiling level K_q of the logistic model.

By choosing specific functional forms for locational benefits and costs, one can thus endogenously determine K_q, the maximum dimension of the region, corresponding to "size" **D** in figure 4.2, where net locational benefits are equal to zero. If, for example, we assume a quadratic formulation for the average net benefits function (i.e. $N_{fq} = an_q^2 + bn_q + c$, with $a < 0$, $b > 0$, and $c > 0$), then the regional equilibrium level **D** would be equal to:

$$\mathbf{D} = \frac{-b + \sqrt{b^2 - 4ac}}{2a} \tag{4.6}$$

By similar calculations every other critical dimension of the cluster can be found[20].

In this first formulation, the number of located firms directly generates (through agglomeration dynamics) the level of locational benefits; since the entry rate is assumed to be proportional to the level of locational benefits, it also indirectly determines the location of new firms into the cluster.

Such a formulation of the location process is very simple but can be used to empirically estimate key parameters of the location path of different clusters[21]. These estimated parameter could also be used as dependent variables in cross-section analyses in order to assess the influence of different factors on the level of the intrinsic growth rate of a cluster or on its maximum dimension.

However this formulation can be criticised on a number of grounds. The first refers to the "isolation hypothesis": firms' location decision are modelled as a dichotomous choice, in the sense that there is only one possible site for location and the choice variable is just the timing of the entry. The second concerns the individual rationality of these firms: the entry rate is modelled as proportional to the level of locational benefits and firms do not anticipate the effects of further locations on such benefits.

---

[19] After **D** new entries thus support a turnover process without causing relevant changes to the equilibrium level.

[20] Referring to figure 4.2 other main critical dimensions — in terms of the underlining parameters of the net benefit function — are as follows: $A=(-b-(b^2-4ac)^{1/2})/2a$, $C=-b/2a$, $X=(-b-(b^2-3ac)^{1/2})/3a$.

[21] Where the difference may refers to different industries in the same geographical site, or to different geographical sites in the same industry.

The following sections are thus devoted to the presentation of two different modelling frameworks which can be used to overcome, at least partially, the main drawbacks of the simple logistic growth path.

4.2.4 Location process in a two-regions framework

Several different interactions may arise when a framework with two regions (1 and 2) is considered. The first concerns the location decisions of a set of industry-specific potential entrant firms confronted with the choice of locating in one out of two regions. This problem can be formally described through a model of inter-regional competition with a given number of potential entrant firms[22]. The entry decision is again modelled as dependent on the number of firms located in each region, reflecting the effects of agglomeration economies and diseconomies, as described in the previous section.

Formally, the multi-region situations can be described through the following system of equations which directly recalls the monoregional development process of equation (4.5a):

$$\begin{cases} \dfrac{dn_1}{dt} = r_1 n_1(t)\left[1 - \dfrac{n_1(t) + c_{12} n_2(t)}{K_1}\right] \\ \dfrac{dn_2}{dt} = r_2 n_2(t)\left[1 - \dfrac{n_2(t) + c_{21} n_1(t)}{K_2}\right] \end{cases} \tag{4.7a}$$

where r_1 and K_1 are respectively region 1 incipient growth rate and equilibrium levels and c_{12}, is the "competition coefficient" which measures the extent to which region *2* compete as an alternative location to region *1*.

However — for expositional purposes — it is more convenient to replace this formulation (which underlines the similarities between the two-regions case and the model of development of one isolated region depicted in 4.5a) with an alternative formalisation which is able to model different multiregional interactions through simple differences in the sign of parameters. We thus replace (4.7a) by (4.7b), where time references are dropped for the sake of clarity:

$$\begin{cases} \dfrac{dn_1}{dt} = \left(a_1 - a_{11} n_1 - a_{12} n_2\right) n_1 \\ \dfrac{dn_2}{dt} = \left(a_2 - a_{22} n_2 - a_{21} n_1\right) n_2 \end{cases} \tag{4.7b}$$

[22] This peculiar approach to the process of multi-regions development (where the region is modelled as an agent and firms as passive industrial mass) reveals its advantages when modelling active policy pursued by different local authorities willing to foster the industrial development of their respective territories.

In system (4.7b), n_1 and n_2 are the "economic masses" (number of incumbents) of each region, a_1 and a_2 are the intrinsic rates of increase of each region in isolation (r_1, r_2 in system 4.7a), a_{11} and a_{22} — the parameters on the quadratic terms which reflect the concavity of the net agglomeration benefits function — i.e. that reflect the inhibiting effects that a firm's entry has on the growth rate of the same region[23] (intra-region competition parameters); a_{12} and a_{21} show the inhibiting effects that one firm — locating in a region — has on the growth of the other region[24] (inter-regions competition parameters).

According to the relative value of these parameters, three main outcomes[25] are possible, as shown by phase diagrams in figure 4.3.

Phase diagrams are useful tools to describe systems of differential equations. At any moment in time the state of the system is fully described by the number of firms located in each region (n_1 and n_2). Within the phase plane one can further draw for each region the locus of points (i.e. combination of n_1 and n_2), which is called the isocline, where the region's industrial mass does not change (i.e. where $\frac{dn_1}{dt}=0$ and $\frac{dn_2}{dt}=0$). The multiple equilibria of the system are thus identified by the intersections of one isocline with either the axes or the other isocline[26]. The arrows in the phase planes indicate the direction in which the system, at that point, will move; the colours and shapes of the dots show whether an equilibrium is either stable or unstable or a saddle point.

---

[23] a_1/a_{11} and a_2/a_{22} in this formulations are the regions' maximum size or isolated equilibrium values (K_1 and K_2).

[24] Different interpretations for these parameters are possible: the simplest relates the values of inter-regional competition parameters to the fact that — given a certain number of potential entrants — each firm that locates in region 1 does not locate in region 2. Then the inter-regions coefficients should be equal, i.e. a_{12}/a_{21}. Another interpretation explicitly relates to the possible differences in the logistic growth path of the two regions in isolation (which reflect differences in the net benefits function). In this case inter-regions competition coefficients can be different.

[25] Selected from the analysis of 13 possible consistent cases. For a more complete analysis, see Capelo (1989, pp. 165-169) and Beltrami (1987, pp. 72-73).

[26] The intersection of an isocline with an axis is an equilibrium point since one region in deserted and the other does not change its industrial mass, the intersection of two isoclines is an equilibrium point since both regions do not register any changes in the number of located firms.

Figure 4.3 Inter-regional competition

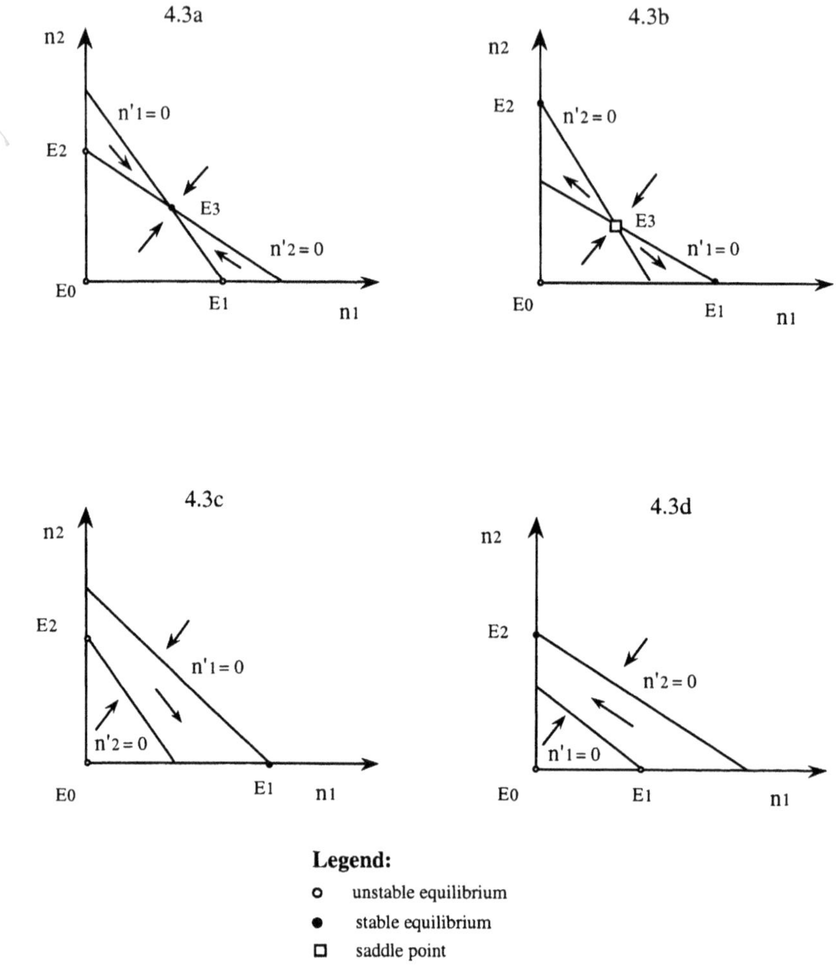

Figure 4.3a shows the persistent coexistence of two regions (both recording a positive number of locations) as a stable equilibrium when:

$$\frac{a_1}{a_{12}} > \frac{a_2}{a_{22}} \quad \text{and} \quad \frac{a_1}{a_{11}} > \frac{a_2}{a_{21}} \tag{4.8a}$$

Figure 4.3b shows the coexistence of two regions as an unstable equilibrium, which is extremely vulnerable to local perturbations and initial biases[27] when:

$$\frac{a_1}{a_{12}} < \frac{a_2}{a_{22}} \quad \text{and} \quad \frac{a_1}{a_{11}} > \frac{a_2}{a_{21}} \tag{4.8b}$$

Figures 4.3c and 4.3d show the total exclusion of one region from the location process that occurs only in the other region (irrespective of the initial state) which reaches its carrying capacity leaving the opponent deserted [28]. In particular, only region 1 is entered when:

$$\frac{a_1}{a_{12}} > \frac{a_2}{a_{22}} \quad \text{and} \quad \frac{a_1}{a_{11}} > \frac{a_2}{a_{21}} \tag{4.8c}$$

while only region 2 is entered when:

$$\frac{a_1}{a_{12}} < \frac{a_2}{a_{22}} \quad \text{and} \quad \frac{a_1}{a_{11}} > \frac{a_2}{a_{21}} \tag{4.8d}$$

Since the industrial growth of the region in isolation follows a logistic path, then the isoclines are linear and the stability of equilibrium depends on the equilibrium point being above the line[29] joining the maximum dimensions of each region in isolation: $\left(\frac{a_1}{a_{11}}, 0\right)$ and $\left(0, \frac{a_2}{a_{22}}\right)$[30].

This presentation of the possible outcomes of firms location processes in a two-regions framework would seem to suggest that inter-region competition leads — almost necessarily — to a situation of locational monopoly, in which only one region is chosen as the preferred site by industry-specific firms while the other remains deserted. Although the economic history of industrial regions displays few cases supporting this result, the coexistence of multiple non-empty regions is a more common phenomenon. This must therefore be caused by factors still missing from the above analysis.

One way to explore the problem assumes the intrinsic growth rates of two regions to be equal (i.e. $a_1 = a_2$). Then the persistent coexistence of both regions is achieved when:

$$a_{11} > a_{21} \quad \text{and} \quad a_{22} > a_{12} \tag{4.8e}$$

These two inequalities state that, to obtain persistent coexistence of both regions, an increase in the numbers of incumbents in either region must inhibit its own growth more than it inhibits the growth of the other (e.g. that the degree of

[27] More precisely the equilibrium defining the coexistence of two regions is a saddle point; any initial state off the stable manifold, or any perturbation that move the system far from it, leads to the "deprivation" of one out of two regions.

[28] This case includes two opposite situations of only region 1 surviving and only region 2 surviving.

[29] Whose equation is $n_1=(1-(n_2/(a_2/a_{22}))(a_1/a_{11})$.

[30] Alternatively, if one expresses the regions size in "equivalent numbers" (i.e. as fractions of their maximum dimension), the equilibrium is stable if the total equivalent number at the joint equilibrium is greater than unity (i.e. greater than the equivalent number at the isolated region's equilibrium).

intra-region competition and congestion must be larger then the degree of inter-regional competition). This result is likely to be achieved if the regions are, at least partially, "attracting" different types of firms or, in other words, if firms have significantly heterogeneous locational preferences.

This observation has been acknowledged within the population ecology literature for a long time under the name of the "Gause principle"[31]. This principle states — in the original biological terms — that in a situation where "two species are resource-limited, [...] only one species can survive on a single resource" (Levin 1970, p. 413). Translated into an economic framework, the Gause "principle of competitive exclusion" states that — if firms are completely homogeneous in their locational preferences — then the most likely outcome is the unanimous choice of one single region (the one ensuring at the start the highest level of geographical benefits). The existence of a plurality of non-empty regions inhabited by firms belonging to the same industrial sector can therefore be explained in terms of the existence of different types of firms which have different locational needs and preferences[32].

The degree of competition between two regions is therefore proportional to the degree of locational homogeneity of the firms population. The same concept can be expressed in a rather different way: the degree of inter-regional competition is inversely proportional to each region's specificity[33]. According to this observation, each region should foster the development of specific regional characteristics in order to develop a "niche" strategy able to limit the degree of inter-regional competition. This observation, which has a general value, has a particular interest for any public authority willing to start a new industrial region in an area where there are already other established regions. The identification and the attraction of a specific subset of firms is the crucial element which determine the success, or the failure, of the regional development project[34].

4.2.5 Co-operative interactions

The analytical representation, used in equation (4.7b) to describe the location decisions of a set of potential entrants between two alternative locations, can be easily encompassed to describe other significant interactions which may arise during the development process of regions and/or industrial clusters. These different relationships can be modelled simply by changing the signs (or values) of the inter-regional coefficients a_{12} and a_{21} in the system of equations (4.7b) as shown in table 4.1.

---

[31] Being formally stated by Gause (1934), even though an earlier and slightly different formulation is ascribed to Volterra (1926).

[32] It is very interesting to compare this result to very similar ones — obtained independently and following a totally different approach by Arthur (1988 and 1990) — reviewed in section 3.11.

[33] Where regional specificity means that each region is the ideal location for a different type of firm.

[34] More on this issue can be found in chapter 7.

Table 4.1 Different types of bilateral interactions

Types of interactions	*Sign and values of interaction coefficients a_{12} and a_{21}*
Competitition	- -
Mutualism	+ +
Commensalism	+ 0
Predation	+ -
Neutralism	0 0
Amensalism	0 -

The regional economics literature records several contributions on predatory relationships (see for example; Dendrinos and Mullally 1985; Maino 1989; Nijkamp and Reggiani 1991). This section therefore concentrates on co-operative relations that — despite their empirical relevance — have been neglected by the main stream of population ecology and, consequently, by the early pioneers of "ecological" regional economics[35]. These relationships describe a set of co-operative interactions where the development of a certain set of firms, such as an industrial cluster in a given nation (or one industry in a given region) is not obtained at the expense of the development of another industry (or cluster) but the two processes enjoy synergistic effects.

Both ecological and economic systems record empirical evidence of mutualistic relationships between two populations over a long period of time. There are plenty of historical cases and empirical evidence of these phenomena: from the development of the textile districts in the mining regions of Britain and Belgium in the 19th century[36], to the synergistic development of an industrial and a residential bordering area[37], from the development of a subcontractors' belt around a main

[35] There are two different reasons why mutualistic and amensalistic relations have been ignored. The first reason is an historical one: both Lotka and Volterra — generally acknowledged as the forerunners of population ecology — limited their analyses to predation and competition. The second reason is a theoretical one: in its purest formulation, a mutualistic model does not have a stable equilibrium and one or both populations undergo "an unbounded exponential growth in an orgy of mutual benefaction" (May 1976, p. 95). A growth model without stable equilibrium seemed of no interest to regional and applied economists. However, more sophisticated models of mutualistic interaction allow for saturation in the magnitudes of reciprocal benefits and obtain — as general result — a stable equilibrium with both populations (usually regions, but they can equally be industries) developing larger than in isolation. It must be said, furthermore, that this equilibrium is less stable — in the sense that perturbations are more slowly damped — than in correspondent systems without mutualistic relations.

[36] Textiles firms, in these locations utilised the female components of the miners' families as source of cheap labour force.

[37] The first demanding workers and offering wages, the second demanding goods and offering labour. This interaction goes well when there are not relevant negative external-

industrial core, to the long timed establishment of a cluster of integrated producers and craftsmen in one specific quarter within a city[38].

In particular, we think that the model analytically described in expression (4.9) could well describe the behaviour of two technologically related industries within the same industrial cluster (such as Computers and office machinery and Electronic components), each receiving positive benefits from the development of the other one[39]. We think that the analysis of inter-industry relations is as important as (if not more than) the analysis of inter-regional competition since one of the main findings of the literature on high-tech clusters, and of the original empirical evidence presented in section 6.3, is that the physical proximity of firms belonging to different but interrelated industries $i = (1, 2)$ act as an engine sustaining the cluster's long run growth[40].

The above remarks can thus be described more formally as follows:

$$\begin{cases} \dfrac{dn_1}{dt} = (a_1 - a_{11}n_1 + a_{12}n_2)n_1 \\ \dfrac{dn_2}{dt} = (a_2 - a_{22}n_2 + a_{21}n_1)n_2 \end{cases} \tag{4.9}$$

In this model each industry in isolation would follow a logistic growth pattern, but the presence of one industry in the same cluster has a positive influence on the rate of growth of the other one. The system's dynamics — which is described through a phase diagram in figure 4.4 — always displays three trivial unstable equilibria (figure 4.4a): the origin (corresponding to the underdevelopment of both industries) and each industry's isolated maximum dimension. However, under certain assumptions concerning the slope of isoclines (figure 4.4b), a fourth stable equilibrium (E_3) emerges which enables both industries to reach a higher level in the cluster. The coexistence of two industries in the same cluster allows both industries to grow larger than would have been possible in the isolated case, but intra-industry competition effects prevent the system from experiencing explosive growth.

ities (such as pollution, noise etc.) spilling from the industrial region to the residential one. In such a case we may well describe the situation as amensalistic.

38 For an example of such clustering dynamics, see Storper and Walker (1989, p. 81) which report an amazing map showing the agglomeration in the London watch and clock making quarter of Clerkenwell, in 1861.

39 However it is also theoretically possible that inter-industries relations, within a cluster, become conflicting because of competition on non industry-specific inputs such as real estates, financial services, fiscal and accounting consultants etc. The prominence of cooperative versus competitive interaction must therefore be empirically assessed.

40 Swann defines such a process as "convergence between technologies" and shows that "single technology clusters, while they might grow faster in the formative stages, did not have the lasting power of diversified clusters" (Swann 1998, p. 64).

Figure 4.4 Inter-industry facultative mutualism

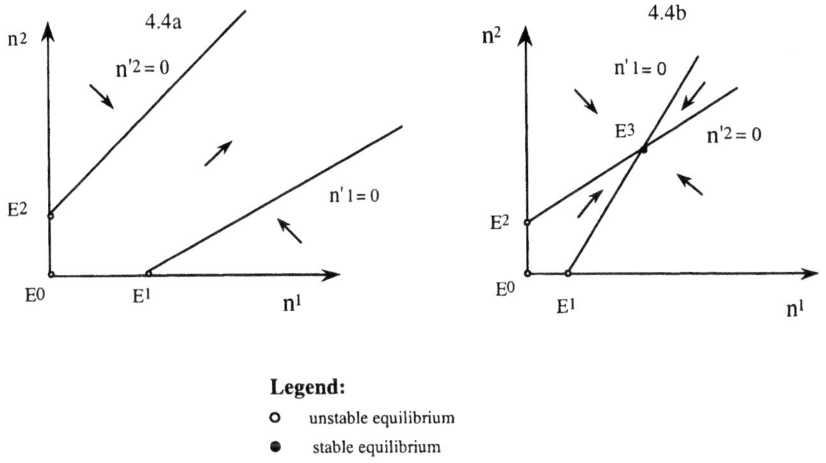

The conditions for an equilibrium solution in which both industries survive can be re-expressed as a strict positivity requirement on the co-ordinates of E_3 in the (n_1, n_2) space, formally:

$$\frac{(a_2 a_{11} + a_1 a_{21})}{(a_{11} a_{22} - a_{12} a_{21})} > 0 \qquad (4.10a)$$

$$\frac{(a_2 a_{12} + a_1 a_{22})}{(a_{11} a_{22} - a_{12} a_{21})} > 0 \qquad (4.10b)$$

Given that all coefficients are positive, the two conditions reduce as follows:

$$a_{11} a_{22} > a_{12} a_{21} \qquad (4.10c)$$

A stable equilibrium in a mutualistic interaction is therefore achieved when the product of the two intra-industry competition coefficients exceeds the product of the inter-industries co-operation coefficients. The same analytical framework may describe the mutualistic growth of two neighbouring clusters, specialised into complementary productions, where the driving forces of the joint development are vertical and horizontal productive linkages established between clusters.

Other stimulating results are obtained when the symmetry assumption required by facultative[41] mutualism is relaxed. An interesting case concerns a situation in which the first cluster in isolation will grow according to a logistic path and its

41 Where the term "facultative" refers to the fact that each industry, in isolation, develops and reaches its equilibrium level. The presence of an industry is therefore no necessary condition for the development of the other one.

growth is further fostered by the presence of industrial activity in the neighbouring one, whilst the second cluster is strictly dependent on the existence of the industrial activity in the first cluster for its very survival. A typical example of this kind is the relationship existing between the "core" of an industrial area (where the large companies are located) and the surrounding area, "periphery", crowded by several small firms (i.e. sub contractors and suppliers) — specialised in accessory productions — which are vitally linked to the industrial success of the core[42]. Similar relationships (which do not imply any spatial differentiation) may well arise between two different industries in the same cluster. A typical example (thoroughly discussed in Swann 1993) is constituted by the relationships between the manufacturing core of an high-tech cluster and the network of specialised service sectors. Since these specific business services must be "consumed" in the very place of their production and cannot be exported to other areas, the existence of a service sector within the cluster is crucially dependent on the existence of a prosperous manufacturing core.

These interactions can be formally described as follows:

$$\begin{cases} \dfrac{dn_1}{dt} = \left(a_1 - a_{11}n_1 + a_{12}n_2\right)n_1 \\ \dfrac{dn_2}{dt} = \left(-a_2 + a_{21}n_1\right)n_2 \end{cases} \tag{4.11}$$

Where n_1 is the economic mass (number of located firms) of the manufacturing core region and n_2 is the economic mass of the specialised service sector.

With the aid of a phase diagram (see figure 4.5), it is possible to see that the system has three equilibria. The first is trivial, unstable and corresponds to the origin; the second, stable, corresponding to the isolated maximum dimension of the core manufacturing industry; the third, which is a saddle point, corresponds to a balanced co-operative relationship[43], in which manufacturing and service sectors coexist with mutual benefits. When the system lies in this unstable mutualistic equilibrium, small positive perturbations lead the system to an unlimited growth while small negative perturbations may cause the "extinction" of the service sector. This is the only possible outcome when $\dfrac{a_2}{a_{21}} < \dfrac{a_1}{a_{11}}$ (see figure 4.5b).

[42] The Motor vehicle industry (which is clustered around Detroit in the USA, Coventry in the UK, Besançon in France, and Torino in Italy) displays a core-periphery spatial configurations in almost every countries.

[43] Which exists if and only if $(a_2/a_{21}) > (a_1/a_{11})$ (see figure 4.5a).

Figure 4.5 Core-periphery dynamics (inter-industry commensalism)

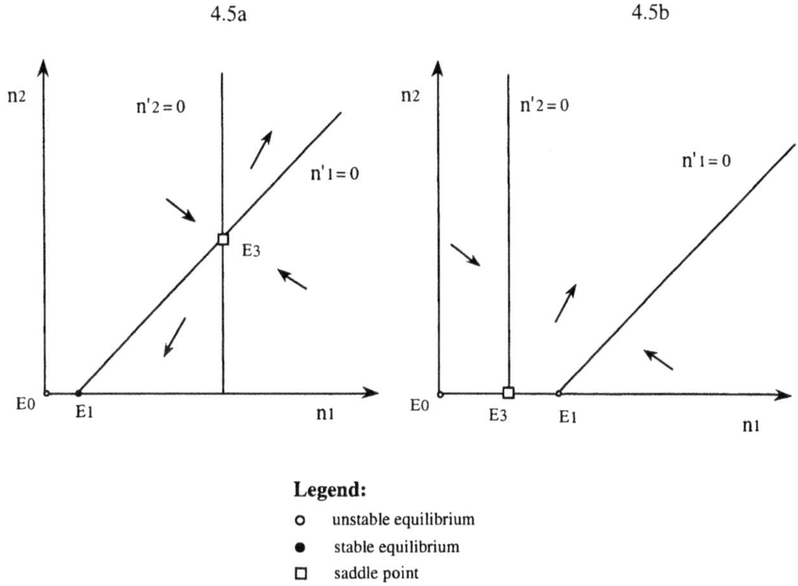

Various type of interactions coexist within the development dynamics of high-tech clusters. The qualitative identification of such interactions and the quantitative measurement of their effect on the growth rate of industries is thus left to the empirical analysis of chapter 6. The effects of the variability of macro-economic conditions on the behaviour of the systems have been extensively discussed in Gambarotto and Maggioni (1998) and are briefly reviewed in section 4.4.

4.2.6 Multi-cluster interactions: some methodological remarks

Previous sections have shown some results that can be obtained by using a series of ecologically-derived models to analyse the growth of a high-tech cluster (or industry) in isolation and of a system of two interacting clusters (or industries).

It would seem therefore natural to extend these results to the case in which more than two clusters are involved. Unfortunately, this extension is not straightforward for empirical as well as theoretical reasons. The firsts refer to the geometrical increase of the relevant interaction parameters, while the second concerns the dynamic behaviour of the system which becomes qualitatively complex.

Two different approaches can be used to overcome this problem. The first relates to the possibility of constructing a large computational model, assigning plausible parameters' values and then simulating the dynamic of the system; the second consists of abandoning the description of individual cluster behaviour —

within a multi-cluster complex — and focussing entirely on some overall aspects of the resulting structure, such as the relation between the increase in system complexity and its overall stability[44].

Even though it is difficult to draw some general conclusions, here are some principles and guideposts for multi-cluster analysis which could be derived from the original population ecology literature (see Levins 1968; Levin 1970; May 1974; Roughgarden 1979) which may give some indications on possible behaviour of such a model.

1. Equilibria may exist only within limited ranges of the interaction and environmental parameters (such as cluster's incipient growth rates, maximum dimension, and macro-economic variability). Outside this restricted range, equations may undergo sustained oscillations and describe a collapsing system in which one or all clusters are completely deserted by firms.
2. Even when the value of parameters ensure an equilibrium solution, this equilibrium may be unstable to large perturbations in the "economic mass" of the cluster. This feature involves the concept of "resilience[45]" of the system and highlights the difference between global and local stability conditions.
3. Fixed parameter values have been assumed throughout the above analysis; on the contrary, in the real world, both interaction and environmental parameters may fluctuate through time, causing further perturbations to the system.
4. Studies based on a generalisation for n-agents of equation (4.7b), in which the coefficients are assigned randomly, show that as the number of agents involved increases, the probability to find a "feasible" equilibrium (i.e. where all clusters register a positive number of located firms) decreases. Biological models also show that dynamic stability typically decreases with the increase in the number of populations, or in the number and strength of inter-populations interactions[46].
5. The previous statement should not be interpreted as a general law stating that, in the real world, complex systems are less stable than simpler one. If a long run perspective is assumed — and we allow the market to act as a selection device — then it is likely that, in a relatively stable environment, a complex and fragile system will persist forever; whilst, in an highly unpredictable and exogenously determined environment the same system will collapse.
6. The Gause principle of competitive exclusion holds for multi-cluster systems as well. In its original ecological formulation it reads as follows: "no stable equilibrium can be attained in an ecological community in which some ε compo-

---

[44] In this section "increased complexity" is, rather loosely, defined in terms of increased number of cluster, more interactions and therefore more parameters. Stability is described as a lower level of fluctuations in the economic mass of the clusters, and as the ability of the system to recover from perturbations.

[45] Which can be defined as a measure of the magnitude of the populations' perturbations that the system will tolerate before collapsing into some qualitatively different dynamic regime.

[46] "Thus, as a mathematical generality, increasing complexity makes for dynamic fragility rather than robustness" (May 1974, p. 160).

nents are limited by less than ε limiting factors" (Levin 1970, p. 419-20)[47]. In economic terms this means that only one cluster can be sustained through a single-industry development which is based on one single type of firms. In other words a system of ε regions is stable only if there are at least ε different types of firms.

4.3 Micro-economic foundation of ecological models: the role of expectations

4.3.1 Introduction

The ecological models presented in section 4.2 give a rather effective description of the macro-economic relationships existing between locational benefits, firms' entry and the growth of an industrial cluster. These models have empirical relevance and are analytically simple. We thus used them as a basis for analysing the development path of an "isolated" cluster and the interactions arising between the developments of several clusters. However these models suffer from one main drawback: namely the total lack of an explicit explanation of individual firm's expectations and behaviour in a dynamic framework. In particular the main problem concerns the difficulty in reconciling a logistic formulation of the entry rate in the cluster implicitly modelled as dependent on the average level of locational net benefits, with an explicit formulation of an individual firm's expectations and behaviour.

From a micro-economic perspective then, as long as: (i) individual firms have even the most primitive rationality, (ii) the presence of incumbent firms signals to outsiders the existence of positive average locational net benefits in the region, and (iii) there are no entry barriers; then (referring to figure 4.2) every outsider will rush to enter until the region has reached the maximum size. Such a process will therefore instantaneously drive the average locational net benefits to zero leaving the region without any appeal for future net entries.

However, this theoretical outcome is in direct contrast with the empirical evidence which shows that the process of firms' locations in a cluster, and consequently the process of regional industrial development, is a gradual and lengthy process (with an average duration of, say, twenty years for high-tech cluster and 50 - 70 years for some traditional industrial districts).

Therefore, from a theoretical perspective, we are left with an open question: why is entry spread over time? Or, in a more formal way, what justifies equation (4.5a)? For answering the questions, it is therefore necessary to identify some sort

[47] Some similarities (not simply formal) can be found between Gause principle and the requirement on the number of objectives not to exceed the number of instruments in the theory of economic policy, and the relation between number of factors and number of commodities in international trade theory.

of mechanism which will control firms' entry (or better, the speed of entry) in the region. This mechanism can either be exogenous or endogenous.

The exogenous mechanism story may run as follows. Either a public agency or some other local authority *de jure* regulates the entry process or — as it is generally assumed by standard epidemic models — the high-tech cluster is initially unknown (as a suitable location) to outsiders and the information on its existence diffuses gradually through the population of potential entrants.

The endogenous mechanism story runs as follows. The time intensity of the entry process necessarily derives from some sort of heterogeneity of firms which affects the actual or perceived locational net benefits available to the single incumbent and determines therefore, different dates of entry for different firms. This heterogeneity can, in turn, be of a twofold nature:

1. objective: i.e. depending on existing differences in the (technological, productive, organisational, financial) structure of firms[48];
2. subjective: i.e. depending on existing differences in the expectations or in the risk aversion of firms.

In reality this sharp distinction may well become blurred, as it seems reasonable to assume that both expectations and risk attitudes of firms (i.e. subjective elements) depend on some objective difference within the firm population.

Given that the literature on the process of inter-firms diffusion of innovation has already extensively discussed the issue of objective heterogeneity (and an explicit application to location process will be presented in section 4.5), we will restrict ourselves to the analysis of the subjective source of firms' heterogeneity.

4.3.2 Subjective heterogeneity: the role of expectations

Assume that all firms, both outsiders and incumbents, are identical as far as technology, size and any other relevant variable are concerned[49] but differently perceive the distribution of locational net benefits. Assume also that the population of outsiders is composed of an infinite number of "types" x_i of firms (each type being characterised by a specific curve of locational net benefits as a function of the number of incumbents in the region).

Let:

$$N_{xq}(n_q) = a_x n_q^2 + b_x n_q + c_x \tag{4.12}$$

where $a_x < 0$; $b_x > 0$; $c_x < 0$; $\left(\frac{-b_x^2}{4a_x} + c_x \right) > 0$

[48] A similar hypothesis is made in Davies (1979) where different adoption dates are explained in terms of: size, management's education and age, self-financing rate, profit rate, growth performance etc.

[49] From this assumption we may derive that firms already located in the region share proportionally both local resources and public infrastructures and therefore that the relevant variable becomes average net locational benefits.

to be the function of expected average locational net benefits of a type-x firms observing n firms already located in cluster q (see fig. 4.6).

Figure 4.6 Expected locational net benefits for type-*x* firm

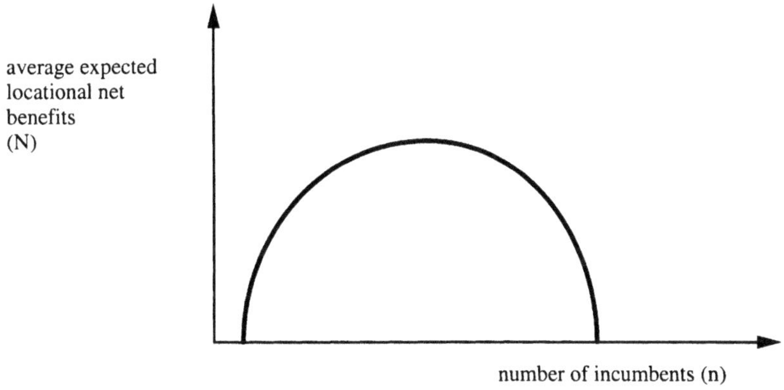

From equation (4.12) it is easy to see that firms expect different locational benefits to be available at different sizes of the industrial mass located in the region. In particular, as already assumed in the previous sections, they expect the relationship between incumbent and locational benefits to be concave and non monotonic (as in section 4.2.2). For the sake of analytical simplicity, and without any loss of generality, we assumed the functional form to be parabolic.

Further assume, for the sake of presentation, that differences between outsiders can be reduced to the position of the x axis. This is equal to assuming that all firms know the general dynamics of the agglomeration/congestion process but do not know the exact initial (critical mass) and final (carrying capacity) dimension of the same process[50].

Different firm types have therefore different expectations regarding the minimum number of firms (entry thresholds) sufficient to generate enough external economies to make entry profitable and the cluster maximum dimension. *Optimistic* outsiders will think that even a small number of incumbents will generate the required amount of external economies, while *pessimistic* outsiders will think that the required externalities could be generated only by a much larger number of incumbents.

50 This last hypothesis ensures that expected benefits curves of all types have the same shape, and that the difference in expectations is described through different positions of these curves on the x axis. The interest in the agglomeration dynamics driving the entry process of firms, and consequently in the regional industrial development, justify this rather heroic assumption. However alternative modelisations of the differences between firms' expectations (i.e. different shapes of the expected benefits functions) would modify the agglomerative dynamics but without changing the basic features.

The location process will then work as follows: in period (t) the outsiders observe the number of firms already located in the cluster. According to their type (and consequently to their expectations) they decide whether to enter in period $(t+1)$, which they do if the number of observed incumbent is larger than their perceived critical mass.

The expectations of potential entrants can thus be represented by a "family" of parabolic functions described by equation (4.12), where each type of entrant is identified by different co-ordinates on the horizontal axis. The family of parabolas can be described through the following restriction parameters values.

$$\frac{-b_x^2}{4a_x}+c_x=Z \qquad \forall x;\ Z\in\Re^+ \qquad (4.13)$$

$$\frac{\sqrt{b_x^2-4a_x c_x}}{a_x}=H \qquad \forall x;\ H\in\Re^+ \qquad (4.14)$$

$$m\leq\frac{-b_x-\sqrt{b_x^2-4a_x c_x}}{2a_x}\leq M \qquad \forall x;\ m, M\in\Re^+ \qquad (4.15)$$

Condition (4.13) requires the maximum value of the expected benefits functions to be identical for all firms. Any firm (irrespective of its type) has the same expectation regarding the maximum level of locational benefit (or profits) it can enjoy in the region.

Condition (4.14) ensures that the distance between the two intercepts on the horizontal axis is equal for all firms. This is to say that the difference between critical mass and carrying capacity is perceived as equal by any firm irrespective of the type it belongs to.

Condition (4.15) defines an interval $(M-m)$ in which the smaller intercept of the locational net benefits must fall and thus strictly defines the family of parabolas. This condition ensures that every type of firm has a finite value for its entry threshold.

From (4.14) we derive that:

$$\sqrt{b_x^2-4a_x c_x}=Ha_x$$

Thus (4.15) can be re-expressed as follows:

$$m+\frac{H}{2}\leq\frac{-b_x}{2a_x}\leq M+\frac{H}{2}$$

Therefore the most optimistic outsider (which needs the smallest number of incumbents to expect sufficient external economies to justify its entry) will have the following minimum intercept on the x axis:

$$\frac{-b_x}{2a_x}=m+\frac{H}{2}$$

while the same value for the most pessimistic outsider will be as follows:

$$\frac{-b_x}{2a_x} = M + \frac{H}{2}$$

The whole family of expected benefits functions is depicted in figure 4.7.

Figure 4.7 Expected locational net benefits for different types of firm

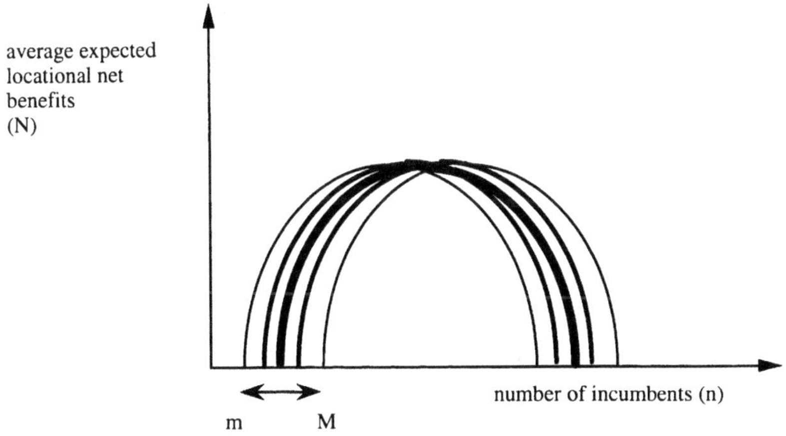

Assume that different firm types are distributed on a density function $f(x)$, with a finite domain $[x_{\min}, x_{\max}]$ which is unknown to potential entrants. Fig. 4.8 shows, as an example, a uni-modal "bell shaped" $f(x)$ function. In this case the population mass distributed on intermediate values is larger than the mass attributed to either extreme values. In other words it is assumed that extremely optimistic and pessimistic firms are a minority of all potential entrants while the larger majority is composed by "intermediate" type of firms.

Figure 4.8 A distribution function of firm's types

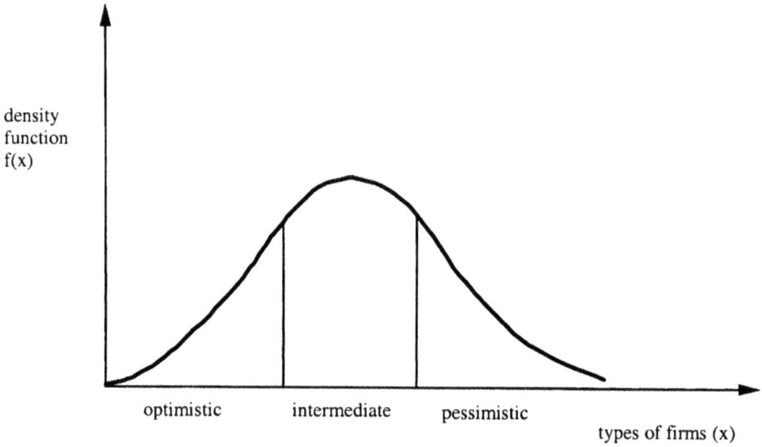

The results presented in this section do not depend in any way on the choice of a particular density function. However different density functions will generate different entry patterns.

Re-state condition (4.15) as (4.16):

$$\frac{-b_{x_{\min}}}{2a_{x_{\min}}}=m+\frac{H}{2}\,; \qquad \frac{-b_{x_{\max}}}{2a_{x_{\max}}}=M+\frac{H}{2} \tag{4.16}$$

and define the function of actual average locational benefits which belongs to the family of parabolas describing the expectations of potential entrants as follows:

$$N(n_q)=an_q^{\,2}+bn_q+c \tag{4.17}$$

where $a_x<0$; $b_x>0$; $c_x<0$; $\left(\frac{-b_x^2}{4a_x}+c_x\right)>0$

This function is identical for every firm, does not have stochastic components and is unknown to the outsider; however it may be reasonable assumed to be known to incumbents which can directly observe the level of locational benefits they are enjoying in a certain time period, given the number of firms located in the region. Although very similar from a formal viewpoint, equations (4.12) and (4.17) are substantially different, since actual and expected benefits usually differ for two main reasons:

1. the expected benefits function of a firm and the actual benefits function are generally different (expectations are generally different from reality);
2. the expectations of any potential entrant for period $t+1$ are based only on the number of incumbents which have been observed in period t. This is due to the hypothesis that outsiders do not know the density function of their types, therefore, they cannot forecast how many outsiders will decide to enter the region in period $t+1$. Once entered, firms observe the actual benefits which are deter-

mined by the number of firms located up to period t plus those entered in period $t+1$.

In synthesis: each outsider observes the number of firms located in the region in period t and decides whether to enter the cluster or not (in period $t+1$) on the basis of expected benefits. It will enter the region if the expected benefits for period $t+1$ are non negative: the *outside option* is assumed to be zero benefits. Once entered, the firm can observe its actual benefits level and, on this basis, decides whether to stay or to exit the region. There are no fixed cost of entry and/or exit[51]. Actual benefits are unobservable by outsiders. Outsiders therefore do not learn anything from incumbents' experiences and continue to base their decision on the expected benefits function.

4.3.3 Entry dynamics

From the assumptions presented in the previous sections it is possible to derive a model of firms' entry dynamics and cluster industrial development[52].

Suppose that the economic activity for all firms is divided into an identical series of time periods and that any locational decision (entry/stay/exit) can be taken only at the beginning of each period. Further assume that in the first period t_0 there is a number n_0 of firms already located in the cluster. Their location — which is treated exogenously in the model and which will be later discussed thoroughly[53] — will cause the entry (in the next period) of all the firms which believe that the presence of n_0 firms in the cluster will generate a level of agglomeration economies so to ensure non negative locational benefits.

Formally, the number n_1 of firms which will enter the cluster at time t_1 can be defined as follows:

$$n_1=\int_{X_1} f(x)dx \qquad \text{where } x_1=\left\{x\in\left[x_{min}, x_{max}\right]: a_x n_0^2+b_x n_0+c_x \geq 0\right\} \qquad (4.18)$$

Thus, in the *second period* there are (n_0+n_1) incumbents firms in the region. They can directly observe their actual locational benefits which, for the above mentioned reasons, are generally different from what they expected. Their decision on whether to stay or to exit the region in the next period is therefore taken on the basis of the actual benefit enjoyed.

If they have experienced non negative benefits in period t_l they will stay in the cluster despite the existing difference between expectations and reality (remember

[51] Alternatively an identical fixed cost for firms relocation (being either entry or exit) can be assumed for all firms without changing any further the model.

[52] Some of these assumptions have been analysed in greater detail in Maggioni and Porro (1994).

[53] Alternatively it can be assumed that the outsiders' distribution of types must include an extreme optimistic type of firms (a "pioneer") which will enter the region even in absence of any previous location.

that the outside option is assumed to be equal to zero)[54]. If they have suffered loss, then they will exit the region. All entry costs are assumed to be sunk cost; therefore they do not enter in the calculation of the net benefits for the incumbent firm which are equal to the difference between locational gross benefits and locational costs.

By recalling the definitions of figure 4.2, we define **A** as the critical mass (i.e. the minimum number of firms needed in order to generate actual non negative locational benefits in the region). More formally:

$$\mathbf{A} = \min\left\{n{:}a\,n_q^2 + bn_q + c \geq 0\right\} \tag{4.19}$$

The agglomeration process, and the regional industrial development therefore come to an end if $(n_0+n_1)< \mathbf{A}$, while it continues until the saturation of the region if the opposite holds: $(n_0+n_1)< \mathbf{A}$.

In the latter case, we can easily calculate the entries in the second period n_2 as:

$$n_2=\int_{X_2} f(x)dx$$

$$\text{where } x_2 = \left\{x \in \left[x_{\min}, x_{\max}\right]: a_x(n_0+n_1)^2 + b_x(n_0+n_1) + c_x \geq 0\right\} \tag{4.20}$$

The entry dynamics will they continue similarly in the following periods.

It is worth repeating that the key assumption in the model is that outsiders cannot observe actual benefits. If this is not the case, outsider can observe the level of benefits enjoyed by incumbents and therefore they can know, right from the beginning, the actual locational benefits offered by the cluster. Therefore if non negative benefits are already enjoyed by some firms in the cluster, in the first period, then entry will immediately take place driving[55] the cluster to its maximum dimension **D**.

To ensure that the entry process, once started by a sufficient number of incumbents, does not come to an early end, it is required that the density mass of outsiders is distributed in such a way that for each $n \geq (n_0+n_1)$ there exists at least one firm willing to enter after having observed the presence of (n_0+n_1) incumbents in the region.

If, on the other hand, the distribution of potential entrants (according to their profitability expectations) is concentrated on extreme values, it may happen that the entry process is abruptly terminated. In this situation, however, one would observe the emergence of positive economic rents for the incumbents, for in the cluster there are enough firms to generate a level of agglomeration economies which would ensure positive benefits (while the outside option of every firm is still zero); nevertheless no other firm would enter the cluster, leaving the incumbents experiencing positive revenues forever.

---

[54] To make the argument manageable we assume firms to be myopic about other firms behaviour even if it is evident that the stream of net agglomeration benefits available in the future is crucially dependent on future entries and exits.

[55] The model does not ensure neither the existence nor the stability of the equilibrium. However it is worth noting that in this very special case, the entry dynamics would be heavily conditioned by the outsiders' learning processes.

This paradoxical situation derives obviously from the assumption of the impossibility for outsiders to observe actual benefits. Otherwise the observation of a number of firms obtaining average revenues continuously larger than those obtainable outside the region would, sooner or later, induce further entries.

4.3.4 The initial critical mass

The distribution of expected locational benefits described by the density function $f(x)$ may, in special cases, reduce the size of the initial critical mass needed to get the entry process started.

If potential entrants knew the exact function of actual benefits, then the critical mass needed to generate for new entries would be equal to **A**. However, some very peculiar distributions of the expected benefits functions may start and foster the entry process even with a critical mass smaller than **A**. In particular this may happen when:

1. there are a number of firms which overstate the extent of agglomeration economies and therefore overstate actual benefits;
2. the density distribution is such that the process of clustering is started right from the beginning in the first period;
3. for each $n \geq (n_0 + n_1)$ there is at least one firm that will enter the region when observing $(n_0 + n_1)$ incumbents[56].

Let us go back to the assumption of the existence of n_0 firms within the region at the beginning of the first period. The exogenous presence of these earlier entrants can be due to the following reasons:

1. these firms have spontaneously developed within the region. (i.e. they are existing traditional firms which recently transformed and moved into high-tech industries);
2. these firms have been "imported"[57] by a local public authority or development agency to start and foster the industrial growth of the cluster;
3. the initial critical mass n_0 is composed partly by indigenous firms, partly by outsider firms imported by a public agency with the scope of generating enough agglomeration economies so to start the clustering process.

The last is probably the most realistic representation of the development of an high-tech cluster where both history (previous industrial culture and tradition) and policy actions (economic public interventions) play complementary role in ensuring the success of an area.[58]

---

[56] Only condition (i) is crucial for the reduction of the required critical mass. The other two conditions are needed in order to ensure the further development of the agglomeration process.

[57] Through some sort of incentive (tax rate reduction, infrastructures provision, public contracts subscription).

[58] In this case, when $n_0 < \mathbf{A}$ the critical mass is smaller than what is required to grant non negative benefits for the locating firms. If some firms locate in the cluster sponsored by a

4.3.5 Firms

In this model all firms (both outsiders and incumbents) are identical but they do not know it. They are equal, so they share the same actual benefits function; however they must ignore this, otherwise, even without observing the incumbents benefits, outsiders could induce that, if an identical firm can survive in the region, then their expectation are too pessimistic and they are underestimating the agglomeration economies available to incumbents. This would accelerate the entry process so to compress it, in the extreme case, to just two periods.

It is worth noting that the assumption of identical firms in the model serves only to simplify the modelling of firms decision-making process and the consequent entry dynamics. If, on the other hand, one supposes that firms were different, then this fact could be described through a whole family of actual benefits (i.e. the benefits which are used by firms to decide whether to stay or to exit the region, once entered) instead of the unique function used here. This fact would cause an increase in the formal complexity of the entry/stay/exit processes and would increase the probability of the entry process being stopped once started, while causing little if any change to the inner mechanics of the firms decision making process.

4.3.6 Exit dynamics

Following the entry sequence described in the previous sections, in period t there will be M_t firms established in the region,

$$M_t = M_{t-1} + n_t$$

where M_{t-1} is the number of firms already located in period $t-1$, while n_t is the number of new entries during period t, more formally:

$$n_t = \int_{X_t} f(x)dx$$

$$\text{where } x_t = \left\{ x \in [x_{\min}, x_{\max}] : a_x M_{t-1}^2 + b_x M_{t-1} + c_x \geq 0 \right\} \qquad (4.22)$$

If $\mathbf{D}$ defines the maximum size of the cluster (i.e. the maximum number of firms the cluster can profitably sustain), formally:

$$\mathbf{D} = \max\left\{ n: an^2 + bn + c \geq 0 \right\} \qquad (4.23)$$

then until $M_t \leq \mathbf{D}$ entry will continue, given that any firm entering the cluster will experience non negative benefits.

---

public authority, then these costs will be covered by the community of firms through some sort of taxes. This fact involves problems of social welfare distortion that will not be dealt with in the thesis. It is also possible that the initial mass n_0 is formed through a co-ordinating action of a group of firms, deciding to enter the region together. However, in this last case, it must be taken into account the possibility of free-riding behaviour between co-ordinating firms.

When $M_t \leq \mathbf{D}$, on the contrary, every firm in the region will experience losses. Now, because the revenues outside the region are equal to zero, some firm will be driven out of the region (or business).

The model in itself cannot forecast in advance which type of firms will leave the region first. However two contrasting alternative hypotheses can be formulated: (i) it is likely that the first firms to exit the region will be those who entered last (i.e. the more pessimistic). This is due to the fact that these firms have gained positive revenues for a very short period of time and, therefore, they are financially weaker and less capable to sustain losses; (ii) it is likely that the first firms to exit the region are those who enter first (i.e. the more optimistic). Optimistic firms are in fact convinced that the number of incumbents that fill the cluster up to its equilibrium level is smaller than expected by the average firm. These firms will be the first to think the cluster agglomeration benefits to be almost totally exploited and will likely exit the region in search of alternative locations[59].

The second situation sounds more reasonable if applied to a stochastic version of the model above. In this case, optimistic firms will observe positive revenues even when, according to their prior beliefs, the cluster should have already reached its maximum dimension. But they will attribute these results to fortuitous shocks introducing stochastic elements in the locational benefit function. However, when all firms in the cluster will start to obtain negative locational benefits because of the continuous process of entry, they would think that their expectations were right and they will be the first leaving the cluster[60].

Exits due to the observation of negative revenues will likely cause oscillatory dynamics around **D**[61]. This dynamics will either be converging or diverging, depending on two phenomena which are not explicitly considered in the model: (i) the level of the exits and the types of firms which will leave the region first; (ii) the learning processes, that can modify firms' expectations, based on the difference between expected and actual benefits. Without going into many details, it is evident that any learning process will dampen the oscillations around the equilibrium level; while, depending on the distribution of firms types in the population of potential entrants, the exit of pessimistic firms may start a divergent oscillatory dynamics.

---

[59] Although empirically appealing (describing the behaviour of a class of "pioneer" firms which look for new interesting locations, settle in there, then leave as soon the competition grows), this second hypothesis — in a deterministic framework — must assume the existence of a plurality of location (in contrast with the bases of the model).

[60] Following the same line of reasoning, it should be the case that these firms would be prone to stay in the cluster in the early phases of its development, despite suffering some losses, because these losses could be easily attributed to negative random shocks.

[61] It is worth noting that oscillatory dynamics around a supposed equilibrium value is the most relevant feature of mature clusters (i.e. which have reached their maximum dimension).

4.3.7 Concluding remarks

Section 4.3 contains an attempt to model the nexus between firms' location decisions and clusters' development dynamics. This nexus has been unjustly neglected by the "orthodox" literature which has always analysed separately the micro-economic location issues and the macro-economic development issues. A further contribution to the oblivion of this nexus comes from the so called "implicit approach" to industrial location (Storey 1985; Keeble 1988; Ciciotti 1993) which underlines that, for the majority of firms, the location is a mere by-product of the formation process. According to this approach, a firm's location decision is implicit and often depends on the place where potential entrepreneurs live and work.

The model described above builds upon the ecologically-derived literature and aims to solve some micro-economic problems and dynamic inconsistencies of the model presented in section 4.2 by explicitly introducing the role of firms' expectations in the location process. Firms have been modelled as myopic agents with limited rationality and their heterogeneity has been assumed as dependent on "subjective" elements (i.e. depending on existing differences in the expectations or in the risk aversion of firms).

Even though this model cannot be empirically tested, one may assume that the subjective differences in expectations and/or risk aversion depend on objective elements such as size, industry etc. If these assumptions can reasonably be justified, then one may substitute the "psychological" distribution of figure 4.8 with an "objective" distribution based on some measurable characteristic and check whether the correspondence between the distribution function of firms' attributes and the cluster's development path confirms or rejects, even if in an indirect way, the model.

4.4 Cluster development policies in variable macro-economic conditions

4.4.1 Introduction

Since the late 1980s the importance of local development policies has been stressed both from a theoretical and an empirical perspective. Academics and public decision makers have acknowledged the central role played by industrial clusters and local systems of production in determining growth potential and the economic success of a country. Even the European Union has shifted the focus of its policy towards the regional and the local levels and is currently targeting lagging regions and declining (both industrial and rural) areas in order to create a more cohesive European Union, achieve a stable long term development path and foster the international competitiveness of Europe (CEC 1991).

If regional and local policies play a relevant role in fostering the industrial and economic development, it is necessary to identify a number of criteria for evaluating different interventions and measuring their efficiency and effectiveness. In a context of increasing global interdependence, these policies cannot however be evaluated without reference to the external macro-economic conditions.

This section presents a model of the development of an industrial cluster which extends and complete the "ecological models" presented in section 4.2 by taking into account the effects of external factors such as the variation of the macro-economic conditions at the national and international level. In this way we are able to discuss the role of economic policies which can foster the industrial development of an area through a number of interventions aimed at supporting the maximum rate of growth industrial mass of the cluster (r-policy) or the carrying capacity of the cluster (K-policy).

Aim of this section is to show that the desirability of a given local development policy (which, in this section, spans from pure K to pure r, passing through intermediate and mixed types) depends heavily upon (i) the development stage reached by the cluster and (ii) the external macro-economic conditions and their variations[62].

[62] This section introduce the discussion of policy issues which are thoroughly discussed and analysed in chapter 7.

4.4.2 Policy interventions for local development

The development path of the cluster, as in section 4.2, is described by the logistic curve: $\frac{dn_q}{dt} = r_q\, n_q(t) \left(1 - \frac{n_q(t)}{K_q}\right)$
which is crucially dependent on the value of two parameters: *r* and *K*.

A pure *r*-type policy is designed to increase the positive externalities which are endogenously generated by the location of a new firm in the region. The intrinsic rate of growth, *r*, expresses the largest possible "attraction and generation" power of a given number of located firms and influences the speed of growth of the cluster. An *r*-type policy explicitly supports the role played by agglomeration economies and knowledge spillovers in the development process of an high-tech cluster through appropriate interventions (such as innovation diffusion supporting policies, start-up incentives, provision of business planning services, diffusion of venture capital activities, etc.).

A pure *K*-type policy is designed to increase the regional "carrying capacity", i.e. the region's ability to sustain a given number of profitable firms. Since the carrying capacity is a function of the local endowment of resources (inputs and infrastructures) and of the average level of use of these resources made by resident firms, then a *K*-type policy aimed at increasing the quantity and/or quality of local inputs and infrastructures, and at raising the efficiency of local firms.

r-type policies generate results in the short run, while *K*-type policy needs a longer time period to be effective. On the other hand, while *r*-type merely influence the speed of development, *K*-type policies are the only ones capable of moving the cluster size from a lower equilibrium level to an higher one, thus ensuring higher sustainable long-run development. *r*-type interventions are suited to be implemented in the early stages of development of an innovative industrial cluster when the main problem is the establishment and survival of an initial core of high-tech firms. *K*-type interventions do their best in the late stages of development of a cluster when competition on inputs and congestion of infrastructures are the main obstacles to the further development of the high-tech cluster.

It should be stressed that a public authority with a given budget available for its local policies faces, at least in the short run, a trade-off between *r*-type and *K*-type policies: a large value of *r* must be "purchased" at the expense of a low *K* and *vice-versa*. This trade-off is explained by the fact that a cluster willing to increase its *r* must support an indiscriminate firms' entry, irrespective of their technological capabilities, which consequently decreases the value of *K*. If the objective of the policy is the increase of the value of *K*, then this is achieved through a strict selection of entrants (based on production and technological efficiency criteria) which results, at least initially, in a reduction of the incipient growth rate *r*.

In addition to these pure-types of policy, there are two other alternatives available to public decision makers willing to foster the economic development of an industrial cluster, these being the mixed and the intermediate policies (whose role will be better explained in section 4.4.5).

Mixed policies involve the implementation of a mix of pure *r* and *K* policies to support the development of a cluster, with this mixture being weighted according to the expected probability (see note 7, below) of different macro-economic conditions. These interventions present a combination of instruments aimed at increasing the cluster rate of growth and of other instruments aimed at raising the technological and organisational level of incumbents firms.

Intermediate policies identify hybrid interventions which lie in the range existing between pure r and pure K policies. These interventions support the development of the region by fostering the creation of agglomeration economies and knowledge spillovers but limiting these interventions within a selected group of firms[63].

4.4.3 The relevance of external macro-economic conditions

The desirability of these two policies is also dependent on the macro-economic conditions (both at the national and at the international level) which can affect the economic performance of the cluster.

The definition of the relevant external environment is crucially dependent on the nature of the economic structure of the cluster and in particular on its degree of openness. It is therefore possible to rank regions according to the geographical level of the relevant economic environment from the "isolated cluster" which operates within local input and output markets, to the "open cluster" whose industrial structure has relevant productive and commercial linkages within the national economy, to the "global cluster" which operates within international markets[64].

In an undisturbed environment (when the relevant macro-economic conditions are good and firm's death rate is "density dependent" i.e. depends only on congestion and competition effects within the cluster) a K-type of policy achieves its best results. Strict entry selection and innovation supporting policy produce a slower development path but a higher equilibrium level of the cluster.

In a disturbed environment (when the relevant macro-economic conditions are poor and firm's death rate is influenced by "density independent" factors which impinge heavily upon the local population of firms even before the cluster achieves its carrying capacity), then the only viable policy is to sustain r.

Let the state of the relevant macro-economic environment be described by E_g - a binomial variable where E_1 stands for a boom and E_2 for a slump period — and Y_h be the specific type of policy (i.e. the position of the implemented policy on a continuum stretching from *pure r* to *pure K* policies) implemented in the cluster.

v_q, the growth rate of cluster q, depends on the combination of Y_h and E_g. Formally:

$$v_q = v_q(E_g, Y_h) \qquad g = 1,2 \text{ and } h \in P \text{ where } P = \{r, K\} \tag{4.24}$$

---

[63] Science parks are the best example of intermediate policies. For more on this issue see sections 7.6 and 7.7.

[64] This taxonomy is thoroughly discussed and empirically implemented in Maggioni and Gambarotto (1997).

Without any loss of generality it can be assumed that v_q — which in the original ecological literature is defined as "fitness function" — approximately follows a normal distribution as a function of Y_h for any type of prevailing macro-economic conditions E_g (see figure 4.9).

Figure 4.9 Growth of the cluster in two different macro-economic environments

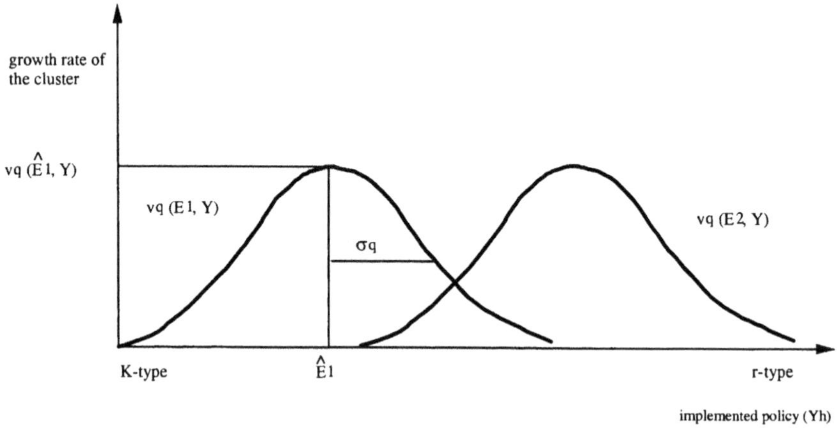

The position of the distribution peak $\hat{E}_g$ on the Y_h axis represents the best development policy to be implemented in the cluster given the prevailing macro-economic conditions, the height of the peak $v_q\left(\hat{E}_g\right)$ measures the best performance (in terms of growth rate of the cluster) which can be obtained by implementing the policy Y_h, and the width of the curve (expressed by the horizontal distance between the peak and the inflection point of the distribution, σ_q) measures the tolerance of the cluster for non-optimal policies.

The fitness function v_q measures the impact of public policies or, in other words, the dependence of the cluster growth performance on policy interventions[65]. Although the height of the peak is undoubtedly limited by the intrinsic characteristics of the firms entering the cluster, it is reasonable to assume the existence of a trade-off between the width and the height of the fitness function so that the width of the function cannot be increased without lowering its height.

The distance between two different macro-economic conditions can be measured by the degree of overlapping of the fitness function; in particular if the near-

[65] A "tall and thin" fitness function (where $vq(\hat{E}1)$ is large and σ_q is small) identifies a cluster which is deeply dependent, and thus very reactive to, on public interventions; a "flat and thick" fitness function (where $vq(\hat{E}1)$ is small and σ_q is large) identifies a cluster which is unresponsive and/or indifferent to public interventions and whose development and growth depend only on its internal forces.

est inflection points of the fitness functions overlap, then the macro-economic conditions can be considered relatively similar; if this is not the case, then the macro-economic condition can be considered to be significantly different from one another.

These two alternatives can be more effectively illustrated through another representation (see figure 4.10) where the fitness function (i.e. the growth rate) of the cluster for each macro economic condition is plotted on each axis. The frontier of the resulting "fitness set", $\varphi_{E_1,E_2}(Y_h)$, synthetically represents the optimal growth rate of the cluster for each state of the world (i.e. the macro-economic conditions plotted on the axes). When the states of the world are significantly different, the set is concave along the north-east boundary; when the states of the world are similar, the fitness set is convex.

If the horizontal distance between the peaks of the fitness functions is greater than twice the tolerance or, more formally, if

$$\left|Y_h\left(\hat{E}_1\right) - Y_h\left(\hat{E}_2\right)\right| > 2\sigma_q \tag{4.25}$$

then the fitness set is concave; if the opposite hold, the fitness set is convex.

4.4.4 Variable macro-economic conditions

In order to study the effects of temporal variations in the macro-economic conditions we need to introduce a concept which, in the original ecological literature, is defined as grain[66] and measures the frequency of temporal variation in the macro-economic environments. In particular when high frequency variations dominate, the series of different macro-economic conditions (e.g. booms and slumps) is said to have a fine grain; if, on the contrary, low frequency variations are the most common, the series is said to have a coarse grain[67].

According to the grain of the series we can define two different types of adaptive functions, $A_f\left(E_1,E_2\right)$ and $A_c\left(E_1,E_2\right)$ where *f* stand for fine and *c* stands for coarse grain, which measure (as monotonic increasing functions of their arguments) the fitness of the cluster for different grains (i.e. different frequency of variation) of variable macro-economic conditions. If, for example, we define as *p* the proportion[68] of macro-economic condition E_1 (booms) and, consequently, $p-1$ the proportion of macro-economic condition E_2 (slumps) in a given time period, then a fine-grained series of macro-economic shocks will be perceived by the cluster industrial mass (i.e. by the incumbents firms) as an average of the two different macro-economic conditions. The cluster adaptive function (i.e. the clus-

---

[66] "Grain refers to the pattern of mixing different environments either in a spatial or temporal distribution" (Hannan and Freeman 1989, p. 106).

[67] High and low frequency are defined in relation to the average life span of a firm within the cluster.

[68] *p* can be defined as a proportion in an *ex-post* perspective or as a probability in an *ex-ante* perspective.

ter's long-run growth) can therefore be represented by a linear combination of the fitness functions in each state of the world, weighted by the duration of each macro-economic condition. Formally:

$$A_f(E_1, E_2) = pv_1 + (1-p)v_2 \quad (4.26)$$

When the series of shocks are coarse-grained, the resulting environment will no longer be perceived as an average of the two macro-economic conditions. In order to grant a stable long-run growth of the cluster the implemented policy must ensure the ability to grow (or at least to survive) in both macro-economic conditions. The cluster adaptive function in this case can therefore be represented by a multiplicative function of the fitness functions in the different environments. Formally:

$$A_c(E_1, E_2) = v_1^p + v_2^{(1-p)} \quad (4.27)$$

The difference between a fine and a coarse series of macroeconomic conditions is made graphically evident in figure 4.10 where the first is represented by a straight line, while the second by a convex curve.

4.4.5 Optimal long run development policies

The optimal long run development policy (i.e. the one ensuring the best growth performance) can be identified by the tangency point between the fitness set and the highest adaptive function (see figure 4.10 and table 4.2). These tangency points indicate the maximum growth rate attainable over the entire period considered.

Figure 4.10 Optimal long run development policies: a graphical representation

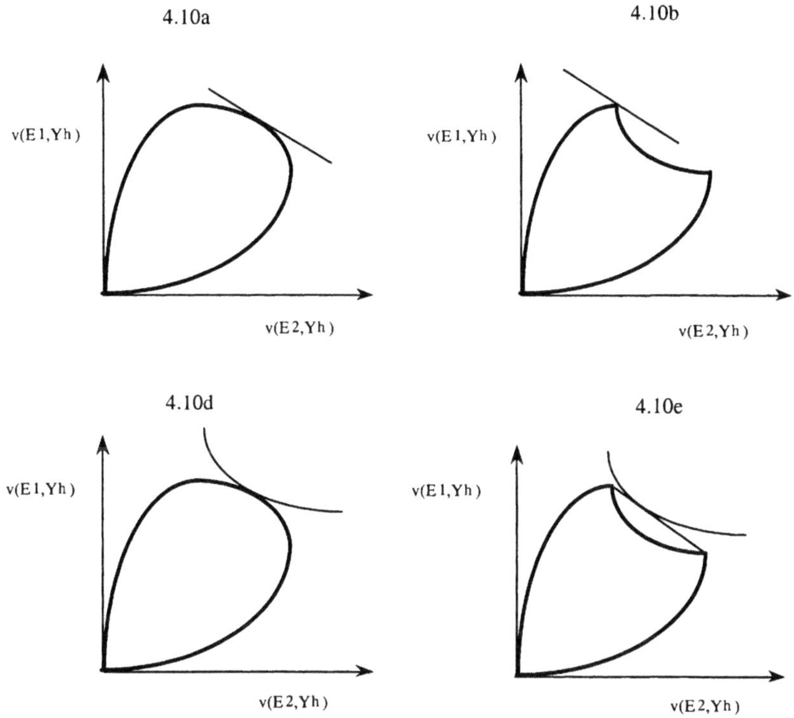

Source: adapted from Hannan and Freeman (1989)

While stable environments obviously favour specialisation — regardless of the grain of the adaptive function (i.e. the frequency of changes in the macro-economic conditions) and of the shape of the fitness set (the difference of two macro-economic conditions) —, variable macro-economic conditions produce more complex effects, which can be better described by table 4.2.

Table 4.2 Optimal long run development policies: a matrix

		difference between macro-economic environments	
		Convex fitness set (the environments are similar)	Concave fitness set (the environments are different)
frequency of shocks	Fine grain (shocks are frequent)	Intermediate policies	Pure policies
	Coarse grain (changes last long)	Intermediate policies	Mixed policies

In particular if the fitness set is convex (i.e. the macro-economic conditions are quite similar) the optimal long run intervention is an intermediate policy between r and K, irrespective of the grain of the series. Intermediate policies are most effective when the depth of exogenous shocks is limited.

If the fitness set is concave (i.e. the macro-economic conditions are very different), then the optimality of the policy is crucially dependent on the prevailing grain: when fine grain prevails, the best development strategy is a pure type of policy; when coarse grain prevails the best development strategy is a mixed policy. When these turbulent conditions prevail, the choice of such eclectic policies diminishes the risk of being locked-in to ineffective development policies and allows, through the change in the relative weights of r and K components, a sharp drift towards pure policies when the macro-economic conditions settle[69].

4.4.6 Concluding remarks

Section 4.4 has shown an application of ecologically derived models to the analysis of local industrial development and local development policies, which emphasises the influence of macro-economic conditions on cluster's performance. This simple analytical framework has not prevented the attainment of significant results. In particular, the above model has clearly pointed out that the success of a development policy for an innovative industrial cluster is heavily dependent on a number of conditions, these being:

1. the development stage of the cluster (i.e. how far is the cluster's industrial mass from its carrying capacity?);
2. the level of dependence of the cluster on policy interventions (i.e. is the development of the cluster heavily dependent on public intervention or is it an autonomous process driven by private investments?);
3. the scale of the relevant macro-economic environment of the cluster, in terms of its productive and commercial relations (i.e. what is the most influential external geo-economic context for the cluster economy; what is the degree of national or international openness of the cluster?);
4. the state of the relevant macro-economic environment (i.e. is the national or global economy in a boom or in a slump?);
5. the depth and frequency of shocks which hit the relevant macro-economic environment (i.e. how different are booms from slumps and how turbulent are the macro-economic conditions?).

---

[69] Maggioni and Gambarotto (1997) contains an empirical application of this modelling framework to regional development in Italy.

4.5 Firm location processes and cluster's development: a micro-economic (innovation diffusion derived) approach

As it has already been signalled in section 3.10, it is possible to consider the process of firms' location, which generates the outset and the development of a spatially defined innovative industrial cluster, as a special case of a diffusion process. The process through which firms choose to adopt a particular innovation does not seem very different from the process through which firms choose to locate in a particular site. The similarity between the dynamics of location and diffusion can be better appreciated if one thinks of a few stylised facts of these two processes.

1. Diffusion and location are time-intensive processes. The saturation of an entire market from the first appearance of a new technology, as well as the saturation of an high-tech cluster from the location of the first high-tech firm, can easily take some years (with the location process being, in general, slower than the diffusion one).
2. When the market penetration of a given technology, and the growth of the industrial mass of an high-tech cluster, are plotted against time, the resulting curve is usually S-shaped, suggesting that diffusion and entry rates are slower at the beginning and at the end of the process, while faster in a middle phase.
3. The speed of diffusion of technology, and the speed of growth of an industrial cluster, display a significant variance across technologies, industries, countries (and regions).

If the location process can be seen as a kind of diffusion process, it is possible to search for a modelling framework — describing the location decision of high-tech firms and the consequent development dynamics of an high-tech cluster — within the vast literature on the diffusion of technological innovation.

Furthermore, models of location process have a strong advantage on the original "technological" counterparts, since they can be conceived of as entirely demand driven on a sound theoretical basis. In location processes, the only rational agent is the potential entrant firm, whilst the supplier of the service/location (i.e. the region) plays a mere passive role[70].

In this appendix we therefore firstly (and briefly) consider a "revised version" of epidemic and probit models and, secondly, we derive a modified version of the encompassing model proposed by Karshenas and Stoneman (1993) in order to look at the different roles played by epidemic, rank, stock and order effects in shaping the development process of an high-tech cluster.

4.5.1 Epidemic models

The basic model of diffusion-location processes is the epidemic model. According to such a model firms, in order to locate in the cluster, must know of its existence

[70] In chapter 7 we relax this assumption and we distinguish between spontaneous clusters and science parks according to the active or passive role of the supply side.

and of the profitability of being located there. The total number of firms located in the cluster at the end of the location process is exogenously determined by the resource endowment (in terms of inputs and infrastructure) of the cluster. The development rate of the cluster is therefore proportional to the probability of interaction between a potential entrant and a firm already located in the cluster.

It is possible to analytically describe such a situation according to an "internal influence"[71] version of an epidemic diffusion model (Mahajan and Peterson 1985). If this is the case, then the entry dynamics in the cluster can be described as follows:

$$\frac{dn_q}{dt} = bn_q(t)(K_q - n(t)) \tag{4.28}$$

where $n_q(t)$ is the number of firms located in the cluster at time t and K_q is the maximum number of firms that the cluster can profitably sustain.

Equation (4.28) states that the speed of the location process is proportional to the frequency of contacts (or social interactions) between incumbents $n_q(t)$ and potential entrants $K_q - n_q(t)$. b is a behavioural parameter which describes the potential entrants' propensity to imitate. For this reason equation (4.28) may well represent a pure imitation location model where potential entrants, once they become aware of the existence of the cluster through the interaction with a firm already located there, decide to imitate the incumbents behaviour and to locate in the cluster[72].

When $n_q(t)$ is plotted against time, equation (4.28) describes the logistic curve, an S-shaped curve, which is often chosen for modelling diffusion processes for its convenient mathematical features[73].

Alternatively, we may assume that the location decision is influenced by some external constant factors a (e.g. incentives to location offered by the local authority). In this case, the time path of the location dynamics become a decaying curve in which, over time, the cumulative number of entrants increases, but at a constant decreasing rate[74]. Formally:

$$\frac{dn_q}{dt} = a(K_q - n_q(t)) \tag{4.29}$$

---

[71] The internal influence model is most appropriate when: (i) the location decision is complex and socially visible, not adopting it places firms at a competitive disadvantage in business; (ii) the set of potential entrants is relatively small and homogeneous; and (iii) there is a need for experiential or legitimising information, prior to location (Mahajan and Peterson 1985).

[72] This formulation of the location process is similar to the ecological model of single-cluster development used in section 3.3.

[73] The logistic curve is symmetrical, has an inflection point at $n(t)/2$, and can be linearly estimated through a logarithmic transformation.

[74] The rationale for such a model is the following: the source of information on the new potential location site is external to the set of firms. Then the rate of location becomes dependent only on the number of potential entrants at each moment of time (and this number is always decreasing as more and more firms decide to locate in the cluster).

where a is a parameter describing the efficiency of the external information spreading mechanisms.

It is also possible to consider these two models in conjunction, allowing both internal and external factors to influence the location process. More formally:

$$\frac{dn}{dt} = \left(a + bn_q(t)\right)\left(K_q - n_q(t)\right) \tag{4.30}$$

These epidemic models can be further enriched by allowing a certain degree of heterogeneity within the set of potential entrant firms.

It can be firstly assumed — by allowing different groups to have different b (as in Davies 1979) — that there are two (or more) distinct types of firms within the set of potential entrants. One is either more anxious to locate in the cluster or more easily influenced by the entrants' decision than the other.

Alternatively, it can be assumed that only the first type of firms is influenced by the external factor a whilst the second type changes its behaviour via direct interactions and therefore is influenced only by the number of firms already located in the cluster through the interaction parameter b (as in Bass 1969; Mahajan and Schoeman 1977).

Another way to look at the location process (as in Mansfield 1968) takes into consideration the differences existing between potential entrants (in terms of location benefits and costs) and explicitly refers to the level of uncertainty which may exist, for the potential entrant, on the profitability of location in the cluster. In this model, it is the uncertainty on the profitable location, and not the mere ignorance of the cluster's existence, which deters potential entrants to locate in the cluster. In the model, firms compare the profitability of the location in the cluster with location (or re-location) costs. Uncertainty is dependent on the date of the first location in the cluster and on the share of incumbents on the total number of potential entrants K_q. In particular it is assumed that as $n_q(t)/K_q$ increases (i.e. as the cluster becomes the best location for a certain industry), uncertainty will be reduced. The proportion of potential entrants which enter the region between t and $t+1$ is therefore modelled as a function of the profitability of location, π; the proportion of firms that have already entered, $n_q(t)/K_q$; and the location (or relocation) costs C.

$$\frac{n_q(t+1) - n_q(t)}{K_q - n(t)} = g\left(\frac{n_q(t)}{K_q}, \pi, C\right) \tag{4.31}$$

The function g is firstly approximated by a Taylor's series expansion with the third and higher order terms ignored as well as the quadratic term in $(n_q(t)/K_q)$, and rearranged in order to write the behavioural parameter b of the original epidemic model (described in equation 4.28) as a linear function of location profits and costs (which can easily be estimated by OLS):

$$b = \beta_1 + \beta_2\pi + \beta_3 C \tag{4.32}$$

4.5.2 Rank models

According to these models, the location process is driven by the differences within the set of potential entrants. Firms differ according to some specific characteristic z (e.g. different location costs) which are distributed across the population as $f(z)$ with cumulative distribution $F(z)$.

At time t a potential entrant f will enter if its characteristic level z_f is smaller than its critical level $\overline{z_t}$ and the proportion of potential entrants which have entered the cluster by time t is equal to $1-F\left(\overline{z_t}\right)$. If one further assumes that, as time passes, $\overline{z_t}$ shifts rightward, the model produces a time intensive location process whose pattern, when plotted against time, is strictly dependent on the shape of the distribution function $f(z)$ (see figure 4.11).

Figure 4.11 Location costs structure and locational gross benefits

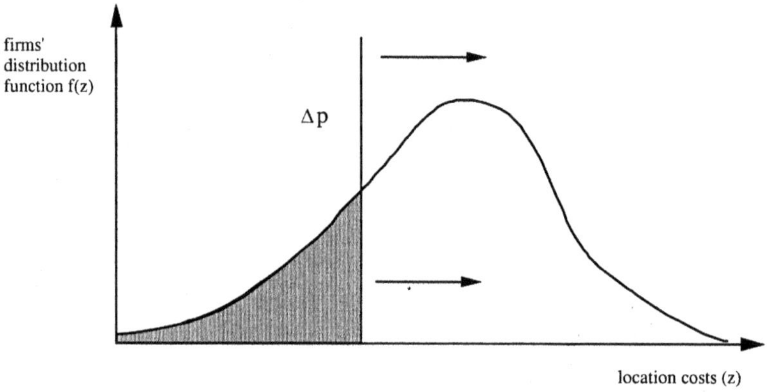

Let's look at the location dynamics in greater details. When a firm of characteristic level z enters the cluster in time t, from that moment onward it will gain a gross locational benefit G_f which is directly related to the agglomeration benefits (e.g. benefits deriving from the location of other firms in term of inputs price reduction, because of the existence of production externalities and technological spillovers)[75].

We can further assume that the price of inputs in cluster q at time t — which depends on the agglomeration economies — is equal for any firm in the cluster. The price of inputs p_q diminishes as the number of firms increases until a threshold

[75] Since we assume that the firm sells its output on a perfect competitive international market, and transport costs are negligible, we can concentrate the analysis on the demand side and model the benefits arising from the location in the cluster as being dependent only on the difference existing between the input prices within the cluster (p_j) and outside the cluster (p_o).

level n_q^m is reached. After that threshold, the relation is inverted (input prices rise, within the cluster, because of congestion and increased competition between firms over a limited supply of inputs[76]). Formally:

$$p_q(t) = p_q(0) - g\left(n_q(t) - n_q^m\right) \qquad \frac{dg}{dn(t)} > 0 \tag{4.33}$$

We can therefore express the locational gross benefits for any firm located in the cluster in time t, $G_q(t)$, as the difference between inputs price within and outside the cluster $\Delta p\left(n_q(t)\right)$.

$$G_q(t) = \Delta p\left(n_q(t)\right) \tag{4.34}$$

Assuming that firm's behaviour is myopic, then a profit maximising firm will enter the cluster if locational gross benefits, deriving from the location in the cluster, are greater than location costs. The development path of the cluster is therefore determined by the structure of the location costs across the population of potential entrants, and by the dynamics of input prices[77].

The dynamic of input prices works as follows. In the first phase, when agglomeration economies are prevailing, the relation between price reduction and firms' entry displays a positive feed-back: as input prices diminish, more firms enter the cluster, and as more firms enter the cluster, input prices diminish. When the number of located firms reaches the threshold level, the relation between entry and price reduction reverses and the process instantaneously stops[78]. If the locational costs structure of potential entrants has a uni-modal distribution (as in figure 4.11), then an S-shaped development path for the cluster emerges. In particular, if location costs are normally distributed, then a logistic path is determined.

An alternative interpretation of the rank model assumes that the relevant difference within potential entrants is psychological. In such models, surveyed in Stoneman (1983) and Thirtle and Ruttan (1987), the population of firms is divided into different psychological classes (pioneers, early movers, early majority, late majority, laggards) and the relative size of these classes makes the overall population to be normally distributed as in figure 4.12.

---

[76] Swann (1998) presents a detailed discussion (and empirical tests) on the different role played by congestion and competition (Cournot) effects.

[77] The relevant difference between this model and the standard probit model used in the technological diffusion literature is that in this model the dynamics of the benefits is endogenously determined by the location of firms in the cluster.

[78] When the entry of the marginal firm decreases $\Delta p(n(t))$, no other location in the cluster can take place.

Figure 4.12 Psychological distribution function of potential entrants

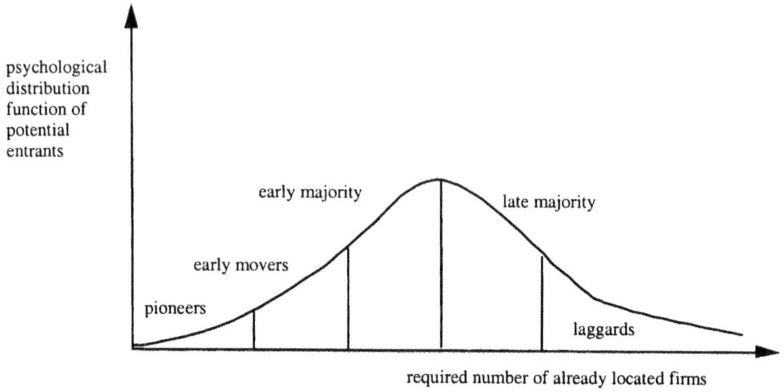

Source: adapted from Stoneman (1983)

What is assumed — by using a normal distribution — is that, generally, within a given population of potential entrants, few firms (pioneers and early movers) will decide to enter a newly developed cluster; a great majority displays a kind of imitative intermediate behaviour; few firms (laggards, or conservatives) wait until the cluster is definitely established as the best location for that industry, before deciding to locate there.

The psychological version of rank models can be further developed by explicitly assuming that the location behaviour of potential entrants crucially depends on previous location decisions (or, which is much the same, on the number of incumbents). In this way the engine of the entry process is made intrinsically endogenous, since present locations are caused by previous locations, and some peculiar dynamic phenomena begin to appear.

A detailed analysis of dynamic systems, and their application to social sciences, can be found in Schelling (1978) and in Hofbauer and Sigmund (1984). A presentation of the major results (in the population ecology modelling framework) is displayed in section 4.3 of the thesis. Here it is worth mentioning that a uni-modal distribution[79] of firm types based on such an endogenous criterion (the distinction between pioneers and laggards is in fact based on the number of previous locations, or threshold, that the two types require in order to locate in the cluster) generates a sigmoid cumulative distribution of firms which, in general, exhibits three dynamic equilibria: two extremes (e' and e''') being stable, whilst the intermediate one (e'') being unstable (see figure 4.13).

[79] As the one described in figure 4.12.

Figure 4.13 The emergence of a critical mass in the development of the cluster

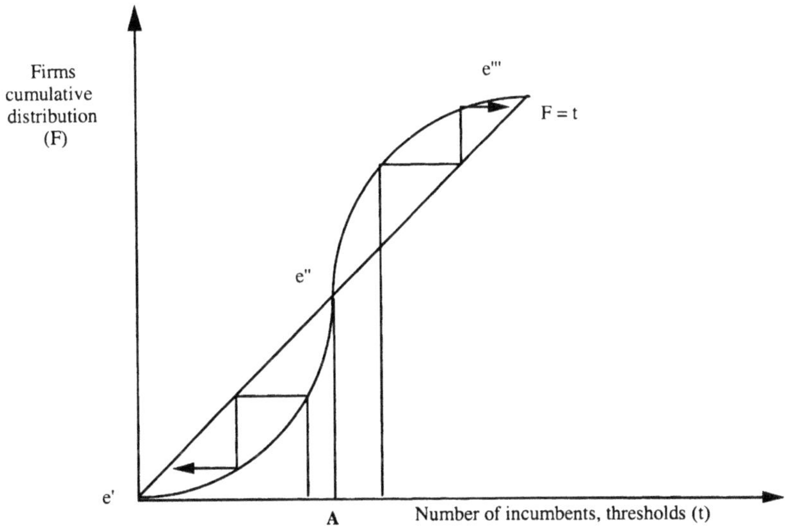

An interesting feature of such a model is the emergence of an endogenously determined "critical mass" (**A**), which is the minimum number of located firms which is necessary for the cluster to sustain its growth. Any number of firms located in the cluster smaller than **A** is doomed to exit since there are not sufficient firms in the set of potential entrants which think it is sufficiently profitable to locate in a cluster which has been chosen by so few firms. A consequence of such a result is that firms will never autonomously start a new cluster and, thus, policy intervention (implemented by an external agent) is required in order to allow the would-be cluster to overcome its critical mass[80].

4.5.3 Rank, stock and order effects in the location process

As already stressed in this chapter, the essential prediction of a location model is that potential entrants have different (preferred) entry dates; or, which is the same, that, at any given date, only a part of the potential entrants actually locate in the cluster. In this encompassing model, derived from Karshenas and Stoneman (1993), we try to show the different role played in the location process by all the different effects[81].

---

[80] The policy implication of these models are discussed in chapter 7.

[81] In order to directly relate to the original source, in this section we use the same notation of Karshenas and Stoneman (1993) and reinterpret it.

Assume that all potential entrants are initially aware of the cluster's existence. Assume also that an outsider firm f can locate in a cluster q in time t by incurring location costs $P(t)$. Define $g_{fq}(\tau)$ as (gross) locational benefits obtained by firm f in period τ from the entry in cluster q. Further assume that these per period benefits are determined by the rank, stock and order effects as follows.

1. Rank effect: firms have different inherent characteristics (size, technology, management abilities etc.) and, as a result, obtain different gross benefits from entering the cluster. Different benefits can thus generate different entry dates. By ranking potential entrants in terms of their returns from entry, a benefit distribution across potential entrants is obtained. A simple entry rule — relating location benefits to entry costs (and depending on the expectation assumptions, changes therein) — enables the derivation of a distribution of reservation entry costs from the benefits distribution. Firms enter the cluster as entry costs fall below reservation entry costs[82].
2. Stock effect: this effect concerns the existence of agglomeration economies and diseconomies. In a static perspective, location benefits enjoyed by the marginal entrant follow a concave non-monotonic path (agglomeration benefits increase until a certain number of entrants is reached, than decrease). In a dynamic perspective, if firms are assumed to last longer than the average life cycle of a cluster and their discount rate is relatively low, then it is optimal for any firm to enter the region as soon as possible to enjoy the stock effects for the longest possible period of time.
3. Order effect: location benefits also depend upon the firm's position in the order of entry, with earlier entrants achieving a greater return than later ones. This can be explained by two different lines of reasoning. The first stresses that early entrants can locate in best geographical sites within the cluster or pre-empt the pool of skilled labour. The second (Fudenberg and Tyrole 1985) refers to the fact that, because of stock effects, earlier entrants get greatest benefits and this induce an entry race[83].

As in the original Karshenas-Stoneman (1993) model, therefore, here order and stock effects push the firms towards earlier entry dates, whilst the rank effect set the earliest possible date for each firm. The only counteracting forces are locational costs, exogenously falling over time, and the discount rate.

Specifically define C_f as a vector of firm characteristics (rank), $K_q(t)$ as the number of firms in cluster q at time t (stock); and $S_q(t)$ as the number of previous entries in the cluster (order). It is evident that, if exits are not allowed, $S_q(t) = K_q(t)$;

[82] Alternatively, rank effects can be modelled as if they could determine only the proportion of the benefits, generated by the other effects (stock and order), that are effectively exploited by the firms. In a sense rank effect can be represented by a parameter a_i (where $0 < a_i < 1$) that acts as a "filter" between the firm and localization benefits. A high ranked firm will enter the cluster also in presence of relatively low locational benefits, whilst a low ranked firm will enter only if the benefits are high.

[83] This explanation emphasises the "first mover advantage" experienced by earlier entrants which can influence the date of entry of their followers, whilst the opposite process is unfeasible.

however, for expositional convenience, it is useful to distinguish between stock and order effects.

The (per period) benefits in time τ for the *f*th firm entering in cluster q at time t will be as follows:

$$g_{fq}(\tau) = g(C_f, S_q(t), K_q(\tau)) \quad \tau \geq t\,;\, g_2 \leq 0\,;\, g_3 \leq 0 \tag{4.35}$$

If r is the discount rate — and assuming no depreciation — then the present value of gross benefits from entry cluster q at time t is:

$$G_f(t) = \int_t^{\infty} g(C_f, S(t), K(\tau))e^{-r(\tau - t)}d\tau \tag{4.36}$$

where the subscript q has been dropped for convenience, since firms are considering only cluster q for future locations.

In this framework, the location decision becomes the choice of an optimal value for t, t^* which is determined by two conditions: the profitability condition, which states that the present value of the net locational benefits $Z_f(t)$ (i.e. the difference between gross locational benefits $G_f(t)$ and costs $P(t)$) at time t must be non negative. Formally:

$$Z_f(t) = G_f(t) - P(t) \geq 0 \tag{4.37}$$

and the arbitrage condition (net location benefits must be not increasing over time):

$$\frac{dy_f(t)}{d(t)} = \left(\frac{dZ_f(t)e^{-rt}}{dt}\right) \leq 0 \tag{4.38}$$

The profitability condition determines, at each moment in time, the set of potential entrants, whilst the arbitrage condition governs optimal entry time, t^*, for each potential entrant.

Under reasonable conditions the arbitrage condition dominates the profitability condition and, therefore, the optimal entry date for firm f, t_f^* is given by:

$$\frac{dy_f}{d(t)}(t_f^*) \leq 0 \tag{4.39}$$

where the inequality allows for corner solutions (i.e. when it is optimal to enter the cluster immediately on its "discovery"[84]).

Assume that firms maximise profits, conjecture that their own decisions do not affect the decision of other firms and use (4.36) and (4.37) to obtain the following expression for $\frac{dy_f}{d(t)}$:

[84] In this context, "discovery" is not to be intended in a strict geographical but in broader economic sense, related to the moment in which a given site become known as a profitable location for a given type of firms.

$$\frac{dy_f(t)}{dt} = rP(t) - \frac{dP(t)}{dt} - g(C_f, S(t), K(\tau)) + \\ + \int_t^{\infty} g_2(C_f, S(t), K(\tau)) \frac{dS(t)}{dt} e^{-r(\tau - t)} d\tau \tag{4.40}$$

where $\frac{dS(t)}{dt}$ and $\frac{dP(t)}{dt}$ measure the expected change in the number of located firms and in the location costs in the small time interval $\{t, t + dt\}$.

Equation (4.40) shows that the benefits for waiting a period before location equal the sum of interests saved ($rP(t)$), plus any expected reduction in the location costs, minus the net present value of the changes in location benefits which results from a move down the entry order for all $\tau \geq t$ (the integral term), and minus the benefits foregone for not being located in the cluster for a period.

Equation (4.40) shows also that $\frac{dy_f(t)}{dt}$, the change in the location benefits deriving from waiting a period before entering the cluster, is positively related to $rP(t)$, $S(t)$, and $K(\tau)$; negatively related to the expected change in the location costs $\frac{dP(t)}{dt}$ and in the number of located firms $\frac{dS(t)}{dt}$; and is also a function of the vector of firm's characteristics C_f, with sign to be determined (since it crucially depends on what characteristics are chosen and how they are measured).

This equation could be used for empirical analysis by making the model stochastic, defining $h_f(t)$ as the hazard rate[85] — or the probability that firm i locates in the cluster in time t, by $(t-1)$ — and utilising the definition that $S_q(t) = K_q(t)$. The final result is as follows:

$$h_q(t) = F\left(rP(t), K(t), C_f, \frac{dP(t)}{dt}, \frac{\frac{dK(t)}{dt}}{r} \right) \tag{4.41}$$

where $F_1 < 0$, $F_2 < 0$, $F_3 \gtrless 0$, $F_4 > 0$, $F_5 > 0$

We therefore expect the probability of location for firm f in period t, given that it has not entered the cluster in the previous periods, to be negatively related to the level of discounted location costs and to the number of incumbents, positively related to the expected change in the location costs and in the number of incumbents. We have *no a-priori* on the sign of the rank effects, since it crucially depends on the chosen feature.

In order to estimate the model, however, we would need data on a series of variables (the level and change in location costs, the individual firm characteristics

---

[85] In Karshenas and Stoneman (1993) the hazard function is chosen so to incorporate in the model the epidemic effects.

— such as size, date of establishment, corporate status — and, above all, the entry date of individual firm) which were not available in the acquired data set[86].

For these reasons, even though we consider the above as a good framework to set up a series of empirical investigations on the process of location[87] — which can capture most of the essential features of the economic theories on cluster growth — we decided to adopt a macro-economic perspective and look at the location process in a modelling framework derived from the population ecology literature. However, in chapter 6, where the empirical analysis is presented and discussed, we succeeded in testing some hypotheses embodied in equation (4.41).

[86] Which does not contain any individual firm level data. It must also be stressed that, to our knowledge, there were not available data series which would contain such information.

[87] And this is a priority in our future research agenda.

Chapter 5 - Survey of empirical literature

> Precisely why particular technologies cluster is not well understood. Innovations often occur in more than one location at about the same time; largely because of different environmental conditions, however, commercial exploitation flourishes in a much smaller number of places. Powerful agglomeration advantages develop in the specialised technopolis. Supplier and service firms arise to serve the growing industry (...). Firms that represent downstream markets join the cluster. A specialised labour market forms, which reinforces the growth of the industry (which, in turn, attracts more specialised labour). Local educational and research institution collaborate with industry to develop programs to meet the need of the industry. (…). In mature technopolises, diversification of the industrial base occurs (as) a natural consequence of the agglomeration process and linkages among certain technologies. (...) The large technical labour force attracts other industries that demand skills similar to those needed by the core industry. (...) Service industries that rose to meet the demand of local industry find export markets and become an independent source of growth for the technopolis. Universities (…) broaden their areas of specialisation and generate new growth in new fields.
>
> R.W. Preer (1992), *The Emergence of Technopolis.*

5.1 Why do firms cluster?

In general, firms aim at maximising profits, therefore firms will choose a location which maximises locational net benefits[1]. These in turn are composed of *geographical benefits* — i.e. benefits deriving from the quality of the site in terms of inputs endowment, costs and consumers location (i.e. the spatial distribution of

[1] Net benefits are obtained as the difference between gross benefits and costs.

demand)[2] — and *agglomeration benefits* (i.e. benefits deriving from the location of other firms in the same site).

If only geographical benefits were considered, then clustering would be the only possible outcome of the location process. For any type (or industrial sector) of firms there would exist an industry-specific "best location" which potentially attract the entire set of firms. Once this "first best" location has reached its carrying capacity[3], other "second best" sites will be chosen. When representing firms as dots on a geographical map, the final picture will thus be a scattered pattern with roughly circular spots marking "good" locations, the density of the spot decreasing steadily as the distance from the spot centre increases, and the size of spots signalling the quality of each location (larger spot, better location).

When agglomeration benefits are added to the picture, the complexity of the situation increases exponentially. Such benefits in fact depend (in a non-linear and non-monotonic way) on other firms' location decisions. This fact introduces two elements of difficulty into the analysis: non linearities and strategic behaviours. As far as non linearities are concerned, one may think of them as being the results of two conflicting forces. On the one hand, agglomeration economies (or agglomeration gross benefits) which produce increasing returns to location in certain sites[4], facilitating the clustering process; on the other hand, agglomeration diseconomies (congestion effects) and inputs competition[5] which push firms apart from each other.

As far as strategic behaviour is concerned, one may see the location process as a game in which each firm must play against a large numbers of, often unknown, rivals in order to find the best site and the best moment to locate in that given site[6].

A rich empirical literature — which is summarised in the following sections — identifies four main reasons why firms cluster:

1. because they want to benefit from agglomeration economies;
2. because they want to benefit from a localised "source" of inputs and/or consumers;
3. because they want to reduce locational search costs and are sensible to informational cascades;
4. because they want to increase their market power.

---

[2] Because the thesis is focused on innovative and high-tech firms — which generally sell their products on a global market with low transportation costs — we will not consider, as relevant, the demand side (market areas etc.), but we will mostly concentrate on the production side of the location problem.

[3] I.e. its maximum dimension either in physical terms: no more land and labour available, or in price terms: because of competition, inputs price has risen over a certain threshold.

[4] And, sometimes, cause locational lock-in phenomena.

[5] The two negative effects can be summarised under the heading of agglomeration costs.

[6] This is not to say that sometimes (i.e. when agglomeration net benefits prevail) location can result in a co-ordination game where each firm benefits from the location of other firms in the same site.

5.1.1 Agglomeration economies

Regional economists, in their jargon, distinguish within the general headings of agglomeration economies between localisation economies (i.e. sector-specific agglomeration economies) and urbanisation/regionalisation[7] economies (i.e. agglomeration economies which are not sector-specific).

- Localisation economies include several elements, such as: *intra*-industry specialisation economies (permitting greater specialisation among firms), labour market economies (reducing search costs for industry specific skills), communications economies (facilitating inter-firm spread of innovations and diffusion of sectoral best practice techniques), scale economies in industry-specific services and intermediate industry-specific goods provision (reducing the individual firm's factors cost). Localisation economies have been empirically measured by using, as a proxy, the local number of firms (or the local level of employment) belonging to a certain industry.
- Urbanisation/regionalisation economies include: *inter*-industry specialisation economies (allowing a higher degree of specialisation among industries and the development of specific production service sectors such as technological auditing, marketing and financial consulting etc.), labour market economies (reducing search costs for generic skills), communications economies (facilitating the cross-fertilisation of innovative activities, the pervasiveness of innovations, and the diffusion of best practice techniques across sectors), scale economies in the production and provision of local public services, goods and infrastructures (reducing general firm's location costs). Urbanisation/regionalisation economies can be seen as external scale economies for a wide variety of economic activities and can be empirically measured by the local level of total employment or by the total number of firms located in a given area.

Agglomeration economies, together with the presence of critical mass phenomena, may also produce the emergence of so called penguin effects[8] (Farrel and Saloner 1986b). This term defines a situation where there is a number of different innovative industrial clusters all in their initial stages of development and each potential entrant is reluctant to move first as long as there is a possibility that his/her choice may turn out to be so inferior as to orphan his/her location decision. He/she may prefer to wait for another user to choose first, in order to free-ride on the informational externalities generated by the location decision.

Agglomeration economies are relevant elements which are taken into account in the location decision problem of every firm (Fujita and Thisse 1996). However, for some specific types of firms (and in particular for small high-tech firms) ag-

[7] The first term, urbanisation, is the most used in the literature and refers to the various benefits, associated with the close proximity to other firms, which arise from the location in an urban area irrespective of the industry. However, following Isard and Schooler (1959), we prefer to add the second term, regionalisation, to stress the point that such benefits can be enjoyed in a spatial environment which is different from a city (such as an innovative industrial cluster).

[8] Which are also known, in the game-theoretical literature, as "wars of attrition".

glomeration economies are even more relevant, since they play a crucial role in determining the site and timing of location.

Small firms need to cluster in order to reach the minimum efficient scale for advertising and marketing (which are then collectively performed), R&D and innovative activities (which are easier in a cluster thanks to informational spill-over), labour recruitment and training (a cluster of firms often produces, directly or indirectly, a local pool of skilled labour), easier access to capital markets (bank managers prefer financing firms locating in established industrial sites, because they can use previous information to assess the profitability of the venture)[9].

High-tech firms cluster to enjoy agglomeration benefits because of their high requirements of information and know-how (higher than those of traditional firms belonging to mature and traditional sectors). The prominence of information in modern economic activities obliges every firm to minimise the costs of acquiring information. However, agglomeration within the innovative sectors, not only minimises the costs of obtaining information, but also allows a firm to maximise the opportunities of acquiring and exchanging it. Firms need also a pool of highly educated and skilled labour (which is usually available where other similar firms are already located) and an established network of backward and forward productive and innovative linkages. Furthermore one should consider that firm spin-off phenomena are very frequent in high-tech sectors and that spin-off, by itself, is a powerful creator of clusters since new entrepreneurs seldom locate their firms far from previous employers (Keeble 1988; Ciciotti 1993).

5.1.2 Localised sources of demand and/or supply

Firm clustering also emerges as the result of a series of independent decisions taken by several firms aiming to locate their establishment in proximity to a particularly well endowed geographical site. The empirical literature on the issue underlines that, for high-tech firms, best location sites can be found within large metropolitan areas close, but not too close, to a great city, near a university and/or research centre, and endowed with efficient transport facilities. Pressure from management and R&D personnel can also influence the location of the firm towards sites with outstanding natural and cultural amenities.

It is difficult to decide whether the location of a skilled labour pool may be classified under this heading because, in theory, labour is a mobile production

[9] However, the fact that location has a primary importance for small firms does not necessarily imply that these firms actually pay a lot of attention to the locational issue or that they are able to select the best location for their needs and to locate there. An empirical analysis (Barkeley and McNamara 1994) — conducted in 1990 over a sample of over 300 US manufacturing firms from Georgia and South Carolina — shows that small firms exhibit little consistency between locational factor ratings and actual firms locations. This seems to suggest at least that these firms perform a very inefficient (if any) locational search process.

factor[10] and it is definitely easier to relocate a household than a plant. Thus, workers should concentrate where firms are located and not vice-versa. If on the contrary labour, and especially skilled labour, is almost immobile, its location becomes one of the main reasons for establishment location and, indeed, clustering.

5.1.3 Search costs reduction, diffusion, and informational cascades

Firm clustering can further emerge because of the explicit decisions of individual firms to follow location decisions of other similar firms, without explicitly taking into account agglomeration economies. Localisation decisions and the regional development process can thus be explained mainly in terms of the economic mass already established in the region, measured through the number (or, alternatively, through employment or total sales) of already located firms. This "going with the stream" behaviour is not at all irrational, for it can be the external manifestation of three different phenomena:

1. imperfect information and search costs: it is often difficult (and/or extremely costly) for outsider firms to observe relevant local variables (such as local factors costs, quality and productivity) needed to forecast future profit streams. The number of located firms (being the joint outcome of two different processes: the location decisions and the effective survival rate in the region) may then be used by outsider firms as the cheapest, easily available proxy for those "hidden" variables. Furthermore, the number of similar firms already located in the area signals the presence of a set of specialised inputs (mainly skilled workers) and the existence of a network of complementary service providers, which can be difficult to assess from outside the region;
2. uncertainty and information diffusion: location decisions in an uncertain environment involve the risk of choosing an unprofitable location. This risk is decreased by observing the behaviour of other firms and/or by establishing relationships with firms already located in the chosen area. Both actions (to observe and to communicate) are directly related to the likelihood of interactions between outsiders and insiders (i.e. potential and actual residents) which is, in turn, positively related to the number of potential entrants and the number of firms already located in the region[11];
3. informational cascades, herd behaviour, and imitation: when economic agents have limited and differentiated information sets, it may be rational to try to improve their decision-making process through the observation of other agents' actions (since actions speak louder than words, the information conveyed by

[10] However in the real world, labour mobility depends heavily on cultural tradition and national institutional frameworks. US labour force is in fact more mobile than UK one (according to Eichengreen 1993, the elasticity of interregional migration with respect to the ratio of local wages on the national average is 25 times higher in the US than in the UK) which, in turn, is definitely more mobile than French or Italian labour force.

[11] This situation is formally depicted in the epidemic diffusion model, where the frequency of interactions (which are assumed to follow a random process) is exactly equal to the product of the number of potential entrants and incumbents.

actions can be more credible than verbal reports in any case). This process can easily lead to the formation of informational cascades causing imitative behaviour even in the absence of payoff interactions, conformity preference or sanctions against deviants (Hirshleifer and Welch 1994). This phenomenon might help in explaining why, when outsiders can see incumbents' actions more easily than their actual revenues, past entries can act as a rational incentive to enter even when personal source of information suggest either to wait or not to enter that particular region.

5.1.4 Increase in market power

Firms can also cluster because of competition. If consumers are evenly spread out over a territory, then firms, according to the corporate geography literature (de Smidt and Wever 1990), can cluster because of two different location strategies: *predation* and *matching*. In the first case entrants try to locate in a particular location — endowed with relevant location benefits[12] — to drive out of business (or, at least in the medium run, out of the site) their incumbent competitors. Thus *predation*, which is a strategy aimed at reducing spatial concentration of firms in the medium term, produces clustering in the short term.

By *matching* strategy we mean that often smaller firms do what larger firms (usually sectoral leaders, which are often first-movers too) have previously done. A matching strategy[13] therefore refers to a leader(s) — follower(s) framework, in which followers react to leaders moves in order to eliminate (or reduce) the leader's emerging competitive advantage based on temporary locational monopoly (Folloni and Maggioni 1994).

More generally one can also remember that — under certain hypotheses regarding the spatial structure of the market and transportation cost functions — clustering can be the simplest results of an Hotelling type oligopolistic location game (the so called minimum differentiation principle).

5.2 How do firms cluster?

Usually firms cluster by industries or by sets of interrelated industrial sectors and create what Isard calls an "industrial complex": "a set of activities (i.e. productive processes) occurring at a given location (...) which are subject to important production, marketing, or other interrelations" (Isard and Schooler 1959, p. 20).

Most of the above mentioned reasons for clustering are in fact relevant only for firms belonging either to the same industry or to some technologically-related industries. In particular, localisation economies and localised sources of inputs im-

[12] This behaviour could also be filed under the previous heading "localised source of demand and/or supply".

[13] Which sometimes is also labelled "catching-up" strategy when the normal industry-specific reaction time-lag is exceeded.

ply a high degree of technological integration and productive disintegration and co-ordination. This process can occur in two distinct but interrelated ways: namely in an horizontal and in a vertical way. Horizontal links define the technological and productive links arising between different firms at the same level of the productive *filière* (these links are the ones ensuring the existence and availability of external economies of scope). Vertical links describe technological and productive links existing between different firms at different levels of the productive *filière* (these links are the ones ensuring a high degree of producer-user interactions, the existence and availability of economies of standardisation and the feasibility of just in time production systems and total quality control processes).

Urbanisation/regionalisation economies, on the other hand, may help to explain why these industry-related clusters often happen to be located near urban centres and/or large metropolitan areas.

The empirical literature on firm location contains a whole set of indicators and factors that have been identified — in different sectors, countries and years — as the most relevant location factors. The following list is the outcome of an extensive review of the literature composed of empirical tests, surveys results (obtained by direct interviews and/or mail questionnaires), econometric analysis, and statistical descriptive studies. Given that each survey ranks these factors in a different order, we decided to list them in the most neutral possible way: by alphabetical order.

1. business climate
2. credit market
   a) efficiency
   b) presence of venture capital
3. governmental influence
   a) public contracts
   b) regulation
   c) taxes and subsidies
4. labour
   a) availability
   b) direct costs
   c) productivity
   d) skills
   e) unionisation
5. material inputs
   a) availability
   b) costs
6. quality of life
   a) cultural amenities
   b) level of educational institution
   c) natural amenities

7. site characteristics
   a) accessibility to information
   b) accessibility to labour
   c) accessibility to markets
   d) housing facilities
   e) geological and climatic features
   f) land price
   g) size of land parcel
   h) transportation infrastructures

8. spatial competition
   a) competitors location
   b) consumers location

9. technological infrastructures
   a) dimension of industrial R&D
   b) dimension of university R&D
   c) number of business service firms
   d) number of firms in related industry

The above list, as complete as possible, gives an account of the many reasons which have been empirically found to determine the location decision of firms. However, being a mere list, it does not give many insights on single issues and on the relationships which exist between different issues. Furthermore, in the above list, location and agglomeration benefits are mixed and tangled. It is therefore worth looking in greater detail at some of these issues in order to assess their empirical relevance.

5.2.1 Relevance of innovative infrastructure

Empirical estimates, based on US data, show that innovative firms actually locate in areas with strong "innovative vocation" or "well developed technological infrastructure" (Feldman and Florida 1994; Feldman 1995). These contributions — based on the analysis of 13 most innovative sectors (ranked according to the number of innovations produced) and using State level data — show that innovative performance is highly correlated with the presence of a sound "technological infrastructure", defined in terms of spatial concentration of (i) firms in related industries, (ii) amount of R&D activity performed by local universities, (iii) local amount of industrial R&D activity, (iv) number of business service firms in the State. Furthermore, Jaffe (1989) and Acs, Audretsch and Feldman (1992 and 1994) note a positive productivity effect for firms associated with the proximity of industrial and academic R&D. This stream of literature[14] aimed at the empirical testing of the R&D spatial spill-over hypothesis, gives a sound economic base to the phenomenon of agglomeration economies for high-tech firms.

---

[14] Which is often quoted, as empirical reference, in the endogenous growth literature.

5.2.2 Relevance of urbanisation/agglomeration economies

Chinitz (1961) contrasts two different metropolitan areas (namely New York and Pittsburgh, one industrially differentiated, the other specialised) to analyse how a local corporate structure might influence the local supply of i) entrepreneurs, ii) capital, iii) labour, iv) land and v) locally produced intermediate goods. He shows that the differentiated city performs better than a "company city". This result can be interpreted as an indirect proof that urbanisation/regionalisation economies can act more strongly than localisation economies. A similar point is made by Jacobs (1969) who compares the development of two UK cities (Manchester and Birmingham) in the nineteenth century[15].

Townroe and Robertson (1980), summarising the results of a plethora of empirical studies conducted on both sides of the Atlantic, conclude that: i) in a highly developed, urbanised and compact economy, local external economies have diminished relevance; ii) such economies remain however important for certain categories of firms and plants (but these sectors and categories remain ill-defined); iii) urbanisation/regionalisation economies are more important than industry-specific localisation economies.

David (1984) and David and Rosenbloom (1990) show the relevance of externalities, arising from the density of labour force congregating at the site and from the size of the capital stock assembled thereupon, in determining the growth and fortunes of a metropolitan area.

5.2.3 Relevance of institutional framework and national characteristics

Geneau de Lamarlière (1991) stresses the point that the same industry can cluster in two radically different ways according to the existence of different institutional frameworks. The case of the semiconductor industry in the US and Japan illustrates how very similar firms decided to privilege access to market (in the first country) versus proximity to production factors, and in particular skilled labour (in the latter) when making their location decisions. Furthermore the concept of "same industry" as defined by industrial statistics classifications may be at risk in international comparisons, due to the wide differences (in market structure, firms dimensions and strategies) existing between different countries. These observations must be therefore considered when making international comparison of development patterns of innovative firms clusters and their respective factors of agglomeration.

---

[15] "At the time of all the intellectual excitement about Manchester, nobody was nominating Birmingham as the city of the future. But as it turned out Manchester was not the city of the future and Birmingham was. (...) Was Manchester, then, really efficient? It was indeed efficient and Birmingham was not. Manchester had acquired the efficiency of a company town. Birmingham had retained something different: a high rate of development work" (Jacobs 1969, pp. 88-89).

5.2.4 Relevance of the availability of skilled labour

Barkeley and McNamara (1994) show that for high-skilled labour intensive firms (a category in which most high-tech companies fall) the most valuable factors when firms are taking location decisions are: "availability of skilled labour" and "quality of education".

The importance of local availability of technical skills and labour for different innovative sectors — such as Aerospace (Bluestone et al. 1981) and Instruments (Oakey 1981a) — has been empirically tested in different countries. Castells (1985 and 1988), Scott (1988a and 1988b), and Twaites and Oakey (1985) stress the point that R&D workers are mobile, because of their relative scarcity, but their mobility is firstly restricted to a subset of geographic locations endowed with natural and cultural amenities, and secondly is limited by social inertia. Markusen et al. (1986) find that high-technology industries are associated with higher wages and higher level of unionisation. Oakey (1981b) underlines that the employment needs of most high-tech firms (which are generally small in size) tend to be incremental and long term. He further stresses that labour cannot be considered as a resource input to the production process in much the same manner as capital, since, in these industries, labour plays a crucial role as a source of new firms formation and, indirectly, as a potent stimulator of innovation (Oakey 1994).

5.2.5 Relevance of venture capital and financial institutions

First developed in the States, in the late 60s and the early 70s, venture capital is the pre-eminent source of start-up funding for over 30% of the high-tech firms located in San Francisco Bay area (Oakey 1984). A study prepared for the Joint Economic Committee of the US Congress (Premus 1984)[16], shows that this particular form of financing has developed strongly following the 1978 capital gains tax reduction and, more interestingly, that this development continued both during the severe 1981-82 recession and the economic recovery of the mid 80s. The geographical distribution of its sources and, above all, of its recipients is strongly biased in favour of four main States — California, New York, New Jersey and Massachusetts — which account for over 75% of the total venture capital deals in 1983 (OTA 1984). Venture capital has taken a long time to be exported to Europe — where it developed mainly in the UK — but the matching of high-tech and venture capital firms still seems very difficult. A survey conducted on 40 UK venture capital firms (Murray and Lott 1995) shows that US venture capital firms invest nearly three times as much into technology based start-ups and early stage investments as their UK counterparts[17]. In Europe, technology based venture capital investments represented 17% of total investments in 1993, with a 20% per-

[16] Based on a mail questionnaire answered by more than 250 leading venture capital firms.

[17] While UK venture capital firms focus on management buy-outs and buy-in and on other later stage refinancing activities.

cent decline relative to 1988[18]. The opposite has happened to venture capital activity in the US, where approximately 80% of venture capital investments financed high-tech firms (Murray 1996).

5.2.6 Relevance of the quality of life

Several authors stress the importance of the local level of "quality of life" for attracting skilled labour to a specific location. Quality of life largely represents urban commercial amenities (the potential for sophisticated leisure and consumption) and other correlates of city cultural and natural resources endowment. Ady (1986), reporting the results of a survey conducted on a sample of 3000 US research engineers, shows the importance of such factors as: housing costs and availability, climate, quality of primary and secondary schools, job opportunities for partner, community attitudes, cultural opportunities, taxes and municipal services. Begg and Cameron (1988) and Keeble and Kelly (1986) show that a high quality of life and the reputation that a place is "high-tech" can outweigh, according to the judgement of surveyed scientists and engineers, the need to locate near major universities. According to Storper and Scott (1989) this does not means that skilled worker preferences determine the location of high-tech firms, but rather than the common interest of mobile professionals[19] and their employers are satisfied best in specific areas.

5.2.7 Relevance of the proximity of R&D plants to head quarters locations

Lund (1986) and Malecki (1979), demonstrate that organisational reasons keep a majority of R&D laboratories and divisions near firm's headquarters. The same point is made by Browning (1980) and Molle et al. (1989), whose analyses show the similarity of the location factors stressed by firms for the two facilities (R&D departments and headquarters). Howells (1984) found that the marked clustering of UK pharmaceutical R&D plants appears to be largely attributable to the location of firms' headquarters (in the South-East area). If one assumes that large firms are early movers in the location process and that imitative behaviours prevail within small firms[20], then he/she can conclude that the location of high-tech industries may be heavily dependent on large firms headquarters prevailing sites.

[18] The difficulties encountered by small innovative firms in obtaining start-up and early stage funding have been acknowledged by the European Commission which, in 1988, has launched a pilot programme (the European Seed Capital Fund Scheme) (Murray 1994).

[19] Buswell (1983) points out the peculiar mobility of high-tech workers. Their relative scarcity gives them labour market mobility; however their locational preferences restrict their geographical mobility to a relatively small subset of places.

[20] As it is suggested by the empirical evidence of US high-tech clusters which shows, over a period of 50 years, a general reduction in the average size of the located establishments.

This pattern of location seems still to be prevailing (especially in Europe) despite the diffusion, since the early 80s, of the Japanese R&D model of "technical branch plants" where R&D is decentralised to actual production sites (Markusen et al. 1986).

5.2.8 Relevance of site reputation, prestige and business climate

The relevance of the above is rarely corroborated by robust empirical results. The most common sources of information are in fact sketchy and rather anecdotal chronicles of the formation and development of famous high-tech clusters (Silicon Valley, Route 128, Research Triangle Park) The analyses listed below are however more rigorous. Entrepreneurial opportunities constitute a strong location factor for R&D workers. The reputation of a site in promoting new firms formation and the number of local spin-off companies act as a strong incentive for highly skilled workers (Malecki 1987; Macdonald 1986; Oakey 1984). The *prestige* of the area is a very important location factors in the case of "artificial" sites (as science parks). Monck et al. (1988) reported the results of an analysis based on a sample composed of 284 British high-tech firms, of which 183 were located in science parks. The main locational factor for this last group of firms was the "prestige and overall image of site and premises". In other studies, extensively surveyed in Malecki (1991, p. 213), the term "business climate" — defined as "a rough metric of a location's expected ability to maintain a productive environment over the foreseeable future" — acts as a sort of composite variable which summarises the length and complexity of some lists of location factors[21]. Sometimes, under the heading of "business climate", one may also find references to the role played by tax rate differentials in determining the location of firms. On this very issues, public decision makers and businessmen on one side, economists on the other, have different and often contrasting, opinions (Moriarty 1980). Empirical estimates on the issues give blurred results, which are heavily industry-dependent. Schmenner (1982) notes that low taxes may be somewhat more valued by high technology industries since they are less locationally constrained by other factors (e.g. access to markets and material inputs). More recently, Papke (1991)[22] shows that, in general, the effective tax rate is negatively correlated with new firm formation but that this correlation loses its significance the higher the technological level of the industry.

[21] It must be stressed that, at least in the US the concept of "business climate", despite its weak scientific basis, is deeply influencing both the location decisions of firms and the territorial marketing of local public government. Annual reports on "States business climates" produced by consultancy companies, have forced local government to promote specific industrial policies in order to improve the State ranking.

[22] Estimating a panel data of 22 US States and five manufacturing industries, from 1975 to 1982.

5.2.9 Excellence and success do not necessarily bring success

This final remark acts as a sort of *caveat* to the previous sections. Feldman (1994) illustrates — through the analysis of the relationship between Johns Hopkins University and the city of Baltimore — that a prominent research university, on its own, sometimes is not enough to create an environment in which high-tech firms can flourish. Further examples come from researchers studying the dynamics of particular high-tech clusters. Using the case history of Boston Route 128 minicomputer industry, Norton (1992) showed that past success can sometime lead to decline and failure. He stresses the point that clustering can be the vehicle of advantages during good times, but also the incubator of epidemics of failures and bankruptcy, because of the close links existing between the local industrial and financial sectors, during bad times. A similar issue is raised by Saxenian (1994) by contrasting the diverging experiences of Silicon Valley and Route 128. The success of the Californian experience and the crisis experienced by the Boston area in the late Eighties cannot be attributed to lack of scientific excellence (Stanford and MIT are both prominent Universities in the most high-tech related scientific fields). The first difference between the two clusters is technological in nature: Silicon Valley exhibits a much more diversified industrial structure than Route 128 (which was, and partly is, still concentrated in computer manufacturing, especially minicomputers). The second reasons relates to the lack of managerial flexibility of firms in Route 128 and their relative inability to adapt to changed technological and market environments[23].

5.3 How to study clustering?

In the surveyed empirical literature there seem to be four main research techniques used to identify the relevant location factors for a given industry (or set of industries): direct surveys (through personal interviews or mailed questionnaires), descriptive statistical studies based on secondary data, econometric analysis, and simulations. This section is therefore devoted to the analysis and discussion of advantages and disadvantages of these methodological approaches.

5.3.1 Advantages and drawbacks of direct surveys

The advantages arising from the use of survey-based data are the following. Surveys are addressed to the person responsible for location decisions and allow researchers to acquire direct information, thus eliminating the need to make infer-

[23] In Saxenian (1994) the lack of flexibility shown by Route 128 is explained in terms of heavy dependence on defence contracts, strong power of professional associations (such as the Massachusetts High Technology Council), and prevailing old-fashioned hierarchical corporate culture. For a contrasting view, based on Route 128 initial comparative advantages, see Dorfman (1983).

ences from secondary data. Surveys also enable researchers to identify minor as well as major factors in location decisions. Surveys results are easy to interpret and report. However, surveys do suffer from a series of drawbacks.

The first is the possible inaccuracy of responses. This may result because individuals may complete the survey in a cognitive dissonant fashion (listing factors believed to be locational attributes but actually not important for the location decision); because firms responses may be sensitive to the design and structure of the survey instrument; because questionnaires may inadequately distinguish between various location factors or omit important factors in the provided list; because the respondent may be less than truthful in the hope of influencing public policy.

The second drawback is related to the cost and the intrinsic features of surveys. Time, difficulty and expenses of the survey may in fact encourage the researcher to excessively limit the geographical and/or industrial scope of the study. The considered sample is generally censored since it is not possible (or, at least, not easy) to survey plants that have been closed or relocated. It is also difficult to determine both the representativeness of the sample and the transferability of findings to other regions or industries.

The third drawback arises from the possible inconsistency between the effective firms' location choices and the preferences they expressed, as it has been shown by Barkely and McNamara (1994).

5.3.2 Advantages and drawbacks of statistical macro-level studies

Statistical studies conducted at a macro level have an undoubted advantage: they are less expensive than surveys both in time and money. Usually such studies are based on already existing data sets which are often produced by national statistical offices[24]. However, these studies suffer from two main drawbacks: a methodological and an operational one.

The first refers to the fact that this research technique focuses on actual firms (or establishments) location. Therefore there is no way to discover the process of locational decision (the considered factors, the set of alternative feasible locations etc.). Data on firms location are then tabulated and confronted (or more simply juxtaposed) with other tables recording local endowment of various geographical benefits. Sometimes simple correlation coefficients are calculated, while very rarely other multivariate statistical analyses (such as principal components, and shift-share) are performed.

The second drawback regards problems which may arise from the differences existing in industrial classifications used in different countries. Other problems may emerge when dealing with data on new firms formation. It is always difficult to discriminate between real "births" and simple legal re-denomination of the same firm, thus causing an overestimation of the turnover process.

---

[24] However, as it has been widely illustrated in chapter 2, to collect and transform data of four different countries, in order to allow international comparison, can be a rather expensive activity both in terms of time and money.

Finally these studies are intrinsically descriptive. The best they can give is a detailed mapping of the clustering phenomenon (which is surely worth looking at) without providing any direct explanatory element.

5.3.3 Advantages and drawbacks of econometric analyses

Econometric analyses are based on the construction of an explanatory model and on its empirical test over a sample of selected observations. From the data gathering perspective this approach shares the same advantages as the previous one. Further difficulties arise when trying to model the actual location process. Firstly there is the problem of how exactly to establish firms' decision time — i.e. the time lag between the collection of the relevant information on the site's features and the date of the firms (re)location — in order to be able to used the correct time-lag structure in the model. Secondly it is often the case that the independent variables (i.e. the locational factors) are highly correlated (thus implying multicollinearity of the observation and inefficiency and instability of the estimated parameters). Thirdly there are problems connected to the censored nature of the variables (usually only data relative to actual located firms are available). Fourthly it is extremely difficult to effectively test a dynamic and strategic formulation of the effect that agglomeration economies play on firms' locations.

5.3.4 Advantages and drawbacks of simulations

Numerical simulations are another way to look at locational dynamics. They are frequently used in order to evaluate complex non linear dynamics, through sensitivity analyses, when analytical solutions cannot be calculated. Simulation techniques can be also used for empirical purposes. In this case the model is fed with past real data and then different solutions (or scenarios) are computed depending on particular values of key parameters. The final results are then compared with real data. Drawbacks of the techniques — when used for empirical interpretation — are connected both with the intrinsic features of the model (i.e. heavy dependence on initial conditions, possibility to display chaotic behaviours etc.) and with the descriptive nature of simulation techniques (which cannot directly investigate the causal relationships between variables but are limited to the analysis of the effects of parameters variation).

5.4 Final remarks

In this thesis we used two out of four above mentioned research techniques: statistical macro-level studies have been conducted in chapter 2 and econometric analyses will be performed in chapter 6.

Aim of the following chapter is to exploit the original database which has been expressly collected for this thesis in order to get empirical answers to the following questions:

1. What are the relevant location factors for attracting high-tech firms and employment in a given area?
2. Are scale economies stronger or weaker than agglomeration economies in determining the industrial specialisation of a cluster?
3. Which is the role of inter-industry linkages, geographical spillovers, and locational shadowing in the process of industrial location?
4. Are there any relevant competitive and synergistic effects in the dynamics of industrial clustering?
5. Which is the more convincing explanation for the growth of an innovative industrial cluster? Arrow-Romer's which stresses the role played by geographical specialisation and monopolistic power, Porter's which underlines the influence of geographical specialisation and fierce competition, or Jacob's which states that local industrial differentiation and competition are the engines of growth?

Chapter 6 - Empirical analyses of the location of high-tech firms and of cluster development

Semiconductor manufacture began in Phoenix (Motorola), and Dallas (Texas Instruments), at about the same time as Shockley Laboratories and Fairchild Semiconductor were established in Santa Clara County in the late 1950s. Aircraft production began in Wichita (Cessna), Buffalo (Curtis), Seattle (Boeing), Los Angeles (Martin, Lockheed, Douglas), as well as in Baltimore and Bridgeport. Farm machinery started up in Stockton (Holt), San Leandro (Holt), and San José (FMC), California, as well as in the Midwest. Yet Santa Clara County, Los Angeles and Illinois become the overwhelming centres of attraction in semiconductors, aircraft and farm machinery, respectively. Only these places developed large complexes of firms producing intermediate inputs as well as final outputs.

M. Storper and R. Walker (1989), *The Capitalist Imperative. Territory, Technology and Industrial Growth.*

6.1 Introduction

The aim of this chapter is to present a collection of empirical analyses which have been performed in order to verify a number of theoretical hypotheses, stylised facts and logical conjectures on the location process of high-tech firms and the development path of high-tech industrial clusters. Because of the multifaceted nature of the issue at study, different empirical exercises have been performed and are summarised in the concluding section of the chapter.

The existing empirical literature has mainly focused its attention on the identification of a list of the most relevant locational factors, while relatively few contributes, which are quoted in this chapter, have analysed the dynamic of clusters' development and its relations with the individual firm's location decision.

Not all models presented in chapter 4 could be directly estimated. In particular the diffusion-derived model of section 4.5 could not be estimated because of the lack of individual firm-level data. We decided therefore to organise the empirical analysis as follows: we first investigate (section 6.2) the role of geographical benefits as determinants of the industrial high-tech specialisation of an area, we then move (section 6.3) to the analysis of scale versus agglomeration economies for explaining the clustering of innovative activities, we further analyse (section 6.4) the effects of inter-industry and inter-regional relations in the dynamic of innovative industrial cluster and, finally, we test (section 6.5) three alternative explanations for the growth of high-tech clusters.

Several reference to the theoretical literature, surveyed in chapter 3, are explicitly quoted in the appropriate sections of this chapter; however it is perhaps worth stressing that section 6.2 displays an empirical analysis in the tradition of classical location theory (and in particular of the least coast approach). Section 6.3 may be considered as an empirical test of both Krugman's and Arthur's approach to industrial location. Section 6.4 builds an encompassing empirical framework able to test some hypotheses put forward by the industrial geography approach, the informational cascades approach and, obviously, by the ecological approach. Section 6.5 is explicitly devoted to discriminate between three alternative explanations: the first which may be ascribed to Krugman, the second to Porter and the third to Jacobs.

6.2 The role of geographical benefits

Even though throughout the thesis locational benefits have been assumed to be composed of two parts — geographical benefits, which derive from the intrinsic quality of the site in terms of inputs and infrastructures endowment and costs, and agglomeration benefits, which derive from the location of other firms in the same site — in most of the modelling framework and in the discussion of policy implications, our emphasis is always on the firm dependent part of locational benefits.

However, many empirical contributions, reviewed in section 5.2, and the practice of day-by-day industrial policy, are focused mainly, if not entirely, on the role played by geographic benefits in influencing firms location decision to accomplish a desired social and economic target (in terms of income, employment, or growth rate of the local economic system).

For this reason we decided to estimate a small number of empirical models which try to assess the relative relevance of different "location factors" in determining the location of high-tech firms and, indirectly, the development of local industrial clusters.

These models generally suffer from a very poor theoretical background. On the one hand, many dependent variables which are used in the regression equations may well be influenced by firms' entry dynamics, thus introducing a problem of endogeneity; on the other hand, such empirical specifications do not take into account the interaction which may exist between the contemporary development of different regions and industries, thus introducing a problem of mispecification.

Nevertheless we used part of our original data-set presented in chapter 2 and supplemented it with other statistical sources in order to relate the location decision of high-tech firms to "objective characteristics" of US States through cross-section and panel data estimations.

The variables used in the analysis are the following:

Dependent variables

$EMPOP_{st}$ = level of employment in high-tech sectors in State *s* at time *t*

$ESTAPOP_{st}$ = number of high-tech establishments in State *s* at time *t*

$GEMPOP_{s}$ = growth rate of employment in high-tech sectors in State *s* (1986-1993)

$GESTAPOP_{s}$ = growth rate of high-tech establishments in State *s* (1986-1993)

$RELEMP_{st}$ = proportion of high-tech employment on total manufacturing employment in State *s* at time *t*

$RELESTA_{st}$ = proportion of high-tech establishments on total number of manufacturing establishment in State *s* at time *t*

Independent variables[1]

$APAY_{st}$ = individual worker's average annual pay in State *s* at time *t*

$UNION_{st}$ = unionisation rate of the labour force in State *s* at time *t*

$UNEMP_{st}$ = unemployment rate in State *s* at time *t*

$MALEPT_{st}$ = male participation rate to labour market in State *s* at time *t*

$ECOL_{st}$ = enrolment rate in college in State *s* at time *t*

TRD_{st} = total amount of R&D expenditure in State *s* at time *t*

PAT_{st} = number of issued patent in State *s* at time *t*

HW_{st} = miles of highway in State *s* at time *t*

---

[1] We collected data on other variables (on female participation in the labour market, number of colleges, university R&D). However, in order to avoid multicollinearity problems, we selected a sub-sample of variables which were more significantly related to the dependent variables. All variables have been used in two different ways: for the cross-section analyses they have been normalised with the population level; while for the panel-data estimations the have been used in absolute levels. For this reason in panel data analyses, population is used as a regressor.

$FAIL_{st}$ = business failure rate in State *s* at time *t*

$INCO_{st}$ = average per capita income in State *s* at time *t*

$EXPO_{st}$ = export rate on total sales in State *s* at time *t*

$METRO_{st}$ = rate of population living in metropolitan areas in State *s* at time *t*

TAX_{st} = corporate income tax marginal rate in State *s* at time *t*

POP_{st} = level of resident population in State *s* at time *t*

6.2.1 Estimating the role of geographical benefits in determining location decisions

The first model which has been estimated reflects the simplest possible explanation for a location process determined by geographical factors. For each time period (1986 and 1993) we run a separate regression.

$$XPOP_s = a + bAPAY_s + cUNION_s + dUNEMP_s + eMALEPT_s + fECOL_s + gTRD_s + \\ + hPAT_s + iHWAY_s + lFAIL_s + mINCO_s + nEXPO_s + oMETRO + pTAX_s + \varepsilon_s \qquad (6.1)$$

where *s* stand for the State, $XPOP_s = ESTAPOP_s$ in models 1 and 3 and $XPOP_s = EMPOP_s$ in model 2 and 4; and ε_s is a white noise error term, assumed well behaved.

The estimated results presented in table 6.1, which are relatively stable for 1986 and 1993, underline the fact that few locational factors are actually relevant for the location of high-tech firms[2].

[2] All usual diagnostic tests have been performed on these rather unstructured and simple regressions, enabling us to reject any significant deviation from the classical model. The use of population as a weighting criterion allows us to eliminate the heteroskedasticity problem which may be generated by the size difference between US States.

Table 6.1 Geographical factors explaining the location decision of US high-tech firms (cross-section analyses for 1986 and 1993)

model	1	2	3	4
year	1986	1986	1993	1993
dependent variable	estapop	empop	estapop	empop
constant	-0.130	14.625	-2.930	-6.149
t-ratio	*-0.08*	*0.48*	*-1.46*	*-0.23*
APAY	-0.006	-0.080	-0.008**	-0.038
t-ratio	*-1.18*	*-0.76*	*-1.77*	*-0.68*
UNION	-0.009*	-0.113	-0.016	-0.093
t-ratio	*-1.89*	*-1.16*	*-1.34*	*-0.61*
UNEMP	0.002	-1.116	0.120	-0.732
t-ratio	*0.04*	*-1.31*	*1.98*	*-0.92*
MALEPT	-0.009	-0.095	0.022	0.135
t-ratio	*-0.34*	*-0.23*	*0.81*	*0.39*
ECOL	10.190	115.398	10.425**	-23.251
t-ratio	*1.57*	*0.90*	*1.67*	*-0.29*
TRD	-0.090	0.204	-0.194	-1.060
t-ratio	*-0.59*	*0.06*	*-1.20*	*-0.50*
PAT	2.0865*	31.642*	2.2445*	21.2408*
t-ratio	*3.39*	*2.59*	*4.02*	*2.91*
HWAY	-0.036	-0.222	-0.049	0.695
t-ratio	*-1.48*	*-0.46*	*-0.98*	*1.06*
FAIL	-0.042	6.894	0.322	32.804
t-ratio	*-0.14*	*1.15*	*1.49*	*11.64*
INCO	-0.010**	-0.003	-0.011**	-0.003
t-ratio	*-1.67*	*-0.02*	*-2.35*	*-0.05*
EXPO	0.015	0.64298*	0.002	0.29482*
t-ratio	*1.18*	*2.58*	*0.15*	*2.09*
TAX	0.006	0.139	-0.009	-0.074
t-ratio	*0.35*	*0.42*	*-0.48*	*-0.31*
METRO	0.006**	0.069	0.005	0.008
t-ratio	*1.83*	*1.10*	*1.44*	*0.16*
Adj. R sq.	0.61	0.45	0.50	0.79
n. of obs.	51	51	51	51

* = 5%, ** = 10% l.o.s.

In particular, when clustering is measured through the number of establishments, the patent variable display a positive and significant coefficient at the 5% l.o.s.[3] and the per-capita income display a negative and significant coefficient;

[3] For the regression on the 1986 data, the unionisation rate displays a significant and negative coefficient showing that high-tech firms, when choosing their location, look for easier industrial relations environments. The rate of metropolitan population displays a positive and significant (at 10% l.o.s.) coefficient too, signalling an influence of urbanisation economies.

while, when the measured variable is the level of employment, patents play the same role and exports, with a positive and significant coefficient, take the place of per-capita income[4]. While the positive correlation with the patent variable has an intuitive explanation — high-tech firms want to locate in highly innovative areas (as measured by the number of per-capita patents) — the negative correlation with the income variable seems rather odd. However, if one think that innovative firms may well look for green-field location far from established production centres (which are still based on traditional industries) and far from metropolitan congestion, then the sign of the correlation may easily be explained.

We tried also to add some temporal dimension to these simple regressions, by using, as dependent variables, firstly the growth rate of $ESTAPOP_{st}$ and $EMPOP_{st}$ ($GESTAPOP_{st}$ and $GEMPOP_{st}$); secondly, by regressing the 1993 values of the dependent variables on the 1986 values of the dependent ones. However these two empirical estimates do not give any valuable results. We decide, therefore, to exploit both the cross-sectoral and the time-series dimension of this data-set and we estimated the following panel data regression equation.

$$RELX_{st} = a + bAPAY_{st} + cUNION_{st} + dUNEMP_{st} + eMALEPT_{st} + fECOL_{st} + \\ + gTRD_{st} + hPAT_{st} + iHWAY_{st} + lFAIL_{st} + mINCO_{st} + nEXPO_{st} + pTAX_{st} + \varepsilon_{st} \tag{6.2}$$

where, following the notation used above, s stands for the State, $RELX_{st} = RELESTA_{st}$ in models 1r and 1f, and $RELX_{st} = RELEMP_{st}$ in model 2r and 2f, and ε_{st} is a white noise error term, assumed well behaved.

We estimated two different regressions, the first where the industrial specialisation of the State is measured in term of establishments and the second where it is measured in term of employment. Both equations have been estimated by considering individual effects to be either random or fixed. Table 6.2 shows the results:

[4] For the regression on 1993 data, the failure rate displays a significant and positive coefficient, showing that dynamic areas where firms turnover rate is higher than average, may encourage firms entry (possibly it is seen as a signal of low exit costs). At 10% l.o.s. the worker's average annual pay displays a significant negative coefficient; while the enrolment in college display a significant and positive coefficient.

Table 6.2 Geographical factors and the location of US high-tech firms (panel data estimation)

models	1r		1f		2r		2f	
	random effects		fixed effects		random effects		fixed effects	
dependent variables	RELESTA		RELESTA		RELEMP		RELEMP	
independent variables	coefficients	*z*	coefficients	*t*	coefficients	*z*	coefficients	*t*
APAY	-1.865	*-0.443*	-11.256**	*-1.895*	-48.864	*-0.691*	-78.8	*-0.415*
UNION	10.435	*1.164*	18.488	*1.194*	-267.731	*-1.439*	9.722	*0.032*
UNEMEP	0.0664	*0.002*	65.483	*1.548*	-4161.789*	*-4.672*	-3294.324*	*-2.435*
MALEPT	15.976	*0.455*	-23.269	*-0.667*	-114.07	*-0.175*	680.644	*0.61*
ECOL	6.625*	*2.037*	8.901*	*2.009*	26.734	*0.587*	120.3	*0.849*
TRD	-0.000805	*-0.01*	0.0399	*0.465*	-0.253	*-0.188*	0.405	*0.147*
PAT	0.796*	*2.206*	0.132	*0.336*	13.126*	*2.095*	5.53	*0.441*
INCO	-0.00138*	*-3.595*	-0.000552	*-0.991*	-0.0158*	*-2.125*	-0.0104	*-0.586*
METRO	0.00528**	*1.649*	-0.00338	*-0.649*	-0.00602	*-0.145*	-0.0203	*-0.122*
EXPO	-4.959	*-0.245*	2.43	*0.108*	991.524*	*3.053*	577.069	*0.805*
TAX	-24.135	*-0.616*	-103.637*	*-2.261*	646.761	*1.075*	128.242	*0.087*
HWAY	1.845	*0.072*	-20.799	*-0.688*	74.402	*0.151*	-28.799	*-0.03*
FAIL	-116.188	*-0.512*	-175.249	*-0.87*	57202.83*	*10.289*	54149.08*	*8.405*
D93	681.922	*2.288*	1173.46*	*3.203*	-10857.77*	*-2.014*	-5218.995	*-0.445*
POP	-0.701*	*-2.40*	-0.208	*-0.35*	-0.341	*-0.086*	-6.928	*-0.365*
constant	1828.086	*0.668*	6693.506*	*2.336*	42974.73	*0.839*	1774.83	*0.019*
Adj. R sq.	0.37		0.37		0.60		0.22	
F test			6.59	*F(15, 36)*			5.86	*F(15, 36)*
Hausman test	15.77	*Chi2(15)*			14.92	*Chi2(15)*		
Breusch-Pagan test	24.91	*Chi2(1)*			18.16	*Chi2(1)*		
number of observations: 102								

* = 5%, ** = 10% l.o.s.

The F tests show that individual effects are relevant, Hausman specification tests indicate that these individual effects should be model as fixed rather than random. As far as the industrial specialisation of States measured in term of high-tech establishments is concerned, when individual effects are treated as random, college enrolment, number of patents (and the dummy variable for 1993, *D93*) show significant and positive coefficients, while population and per-capita income have significant but negative coefficients confirming that the presence of an educated labour force and innovative structures are important location factors for high-tech firms. When individual effects are treated as fixed, population becomes insignificant (since this source of inter-state variance is captured by the model's specification), the quality of the local labour force stays significant and positive and the tax coefficient appears to be significant with a negative coefficient. The worker's average annual pay records a negative coefficient (which is significant at the 10% l.o.s.)

When high-tech employment is the dependent variable for detecting State industrial specialisation (using random effects), the number of issued patents, the export rate and business failure rate show significant and positive coefficients, while the unemployment rate has a negative coefficient. The use of a fixed effects specification results in only two significant coefficients (business failure and unemployment rate). These results indicate that an innovative, dynamic and internationally competitive environment is the best location for high-tech industries, while declining industrial areas are the worst.

6.3 Clustering of firms or workers? Scale vs. agglomeration economies

Recent papers (e.g. Brülhart and Thorstensson 1996; Davis and Weinstein 1996; Dumais et al. 1997; Kim 1997; Ellison and Glaeser 1997; Rombaldoni and Zazzaro 1997; Greenaway and Torstensson 1998) have thoroughly discussed the phenomenon of firms location and regional specialisation both from a theoretical and an empirical perspective.

Classical location theory considers the firm's location decision as a spatial optimisation problem where the spatial distribution of inputs is considered as given and the only strategic element refers to the behaviour of the other firms. Recent theoretical contributions have instead shown that firms' locations and, consequently, regional specialisation patterns are caused by the interplay of the location decisions of firms and workers. In other words if classical location theory can be summarised in the claim "geography matters" — in the sense that the exogenous spatial distribution of inputs (and, sometimes, consumers) crucially determines firms' location decisions —; more recent approaches seem to state that "history, and expectations", matter most.

If this is the case, then the explanation of firms location decision could be found, without referring to exogenously determined locational factors, within the actual firms locational patterns. Krugman (1991a and 1991b) — referring explicitly to Marshall (1920) — stresses the role of economies of scale (which are internal to the individual firm) as the main centripetal force determining firms'[5] location; while other authors (Scott 1986; Arthur 1990; Becattini 1998; Storper and Walker 1984) — quoting almost the same passages from Marshall (1920) — identify agglomeration economies (which are external to the individual firm) as the key determinants of industrial clustering.

The contrast is extended also to the interactions between scale and agglomeration economies. According to Krugman, economies of scale are a pre-condition to the existence of agglomeration economies[6] (thus these two factors coexist and, in general, they are mutually re-enforcing). On the contrary, according to Scott (1986) the very trade-off which exists between agglomeration and scale economies can explain why, in certain industrial sectors and in certain areas, large firms

---

[5] The other centrifugal forces in his models are transport costs and the share of immobile agricultural workers.

[6] "If each firm could produce in both locations (...), then the full portfolio of firms and workers could be replicated in each location and the motivation for localisation would be gone" (Krugman 1991a, p. 40-41).

prevail; while in other industries and/or locations, small interdependent firms seem to be the general rule[7].

Thus the recent empirical literature (Henderson 1994; Kim 1995 and 1997; Hanson 1996; Brülhart and Thorstensson 1996; Ellison and Glaeser 1997, Geroski et al. 1998 von Hagen and Hammond 1998) has mainly tested the relative importance (among other factors[8]) of scale and agglomeration economies in determining the existence of increasing returns to locations and, indirectly, the emergence of industrial clusters.

Some of these issues have been already raised in the second chapter of the thesis where it has been extensively shown that the use of different variables (number of firms versus employment) for measuring local industrial specialisation produces different results. It is however interesting to analyse in greater detail the role played by internal and external economies in determining the geographical distribution of high-tech industries.

6.3.1 A rough indicator

A first technique implies the use of a rough indicator of the relevance of agglomeration versus scale economies which, for each sector, is calculated as the difference between the value of the concentration or inequality indexes referring to establishments and those referring to employment. When the difference is positive (i.e. the sector is more spatially concentrated in terms of establishment) we may say that agglomeration economies are prevailing; when the difference is negative (i.e. the sectors displays a higher concentration index on employment than on establishments) scale economies may well be the stronger engine of clustering dynamics. This procedure can be implemented for each inequality and concentration index. We have therefore built a general index for each industry, as shown in tables 6.3 and 6.4, adding up the values of all indexes.

Table 6.3 Agglomeration versus scale economies at FLA level

Industries	*USA*	*UK*	*FRA*	*ITA*
170	-1.39		-0.67	**0.21**
244	-0.70	-0.06	-0.77	-1.25
300	-0.19	-0.12	-0.86	-2.72
321	0.00	**0.03**	-1.29	-0.40
331	-0.20	**0.02**	-	-0.65
332	-0.22	-0.03	-	-1.03
333	-0.44	**0.01**	-	-0.21
334	-2.33	-0.09	-	-1.62
330	-	-	-0.67	-
340	-1.36	-	-0.78	-2.77
353	-0.82	-0.08	-0.73	-1.79
total h-t	-0.07	**0.02**	**0.09**	-1.81
D	**0.15**	**0.00**	-0.95	-0.40

[7] "Vertical disintegration encourages agglomeration and agglomeration encourages vertical disintegration" (Scott 1986, p. 224).

[8] Kim (1997), for example, stresses the relevance of the location of raw materials and natural advantages.

Table 6.4 Agglomeration versus scale economies at SLA level

Industries	USA	UK	FRA	ITA
170	-1.53	-	-0.01	-0.01
244	-0.63	**0.01**	-0.07	-0.10
300	-0.04	-0.05	-0.13	-0.32
321	**0.10**	**0.17**	-0.03	**0.02**
331	**0.05**	**0.05**	-	-0.11
332	**0.03**	-0.02	-	-0.09
333	**0.09**	**0.36**	-	**0.25**
334	-1.47	**0.04**	-	-0.16
330	-	-	**0.05**	-
340	-1.57	-	**0.04**	-0.14
353	-0.70	**0.12**	-0.03	-0.08
total h-t	**0.03**	**0.02**	-0.03	-0.13
D	-0.12	**0.08**	**0.06**	-0.06

According to these tables, agglomeration economies seem to play a major role at SLA than at FLA level. However some industry-specific features can be summarised as follows. Agglomeration economies play a major role in the Electronic components industry (321), and in the Instruments sector (330) (with specific reference to Medical and surgical instruments (331), and to Industrial process control instruments (333)); while scale economies are prevailing in the Textiles industry (170) (the only notable exception being Italy, at the FLA level because of the relevance of "industrial districts"), Motor vehicles (340) (here is France at the SLA level which display a surprising result), and Optical and photographic equipment (334). When the high-tech sectors are considered as one single industry the results are somehow blurred and seem to be more influenced by national and institutional factors. These tables in fact show some interesting country-specific differences. In particular it seems that, while in some countries (such as the US and the UK[9]), a prominent role in high-tech sectors is played by small firms, in other countries (such as France and Italy), high-tech sectors are still characterised by large size, sometimes publicly owned, firms. The relevance of these results must however not be overstated, since inter-countries differences may have been exacerbated by the rough aggregation of different inequality and concentration indexes.

It is therefore useful to follow the works of Kim (1995 and 1997) and Rombaldoni and Zazzaro (1997), integrating and modifying the approach used by these authors, in order to take into account three other phenomena which play a crucial role in shaping the development of high-tech clusters: the existence of non linearities (i.e. the existence of diseconomies), the presence of inter-industry linkages, and the role of geographical spillovers and locational shadowing.

[9] Even if it must be recalled that, because of confidentiality reasons (as explained in section 2.3.2) employment data for UK were almost unavailable. Therefore the index for UK has been constructed only on the basis of the spatial concentration ratio which, in general, seems to overestimate the effects of agglomeration economies.

6.3.2 A model

The aim of this empirical exercise is to test the relative importance of scale versus agglomeration economies in determining industrial clusters. What we want to test is a general relation which explains the spatial concentration of an industry *i* in region *r* in terms of scale, agglomeration and other relevant variables. Formally

$$spatial\ concentration_{ir} = f\left(scale_{ir}, aggl._{ir}, other_{is}\right) \qquad (6.3)$$

In principle, there are no *a-priori* reasons for choosing one particular functional form for the regression equation. However if one refers back to section 6.3, where the issue of whether these two forces are complementary or substitutes is discussed, it becomes clear that, by choosing a simple additive functional form, we implicitly assume that agglomeration and scale economies are substitutes. On the other hand, by choosing a multiplicative functional form, we implicitly assume that they are complementary. Even though we had an *a-priori* weak preference for the complementarity version of the story, we let the data speak for themselves and we ended up choosing a multiplicative functional form[10] which gave better results on the Ramsey RESET test for functional mispecification.

We enlarged the original model of Rambaldoni and Zazzaro in order to include other relevant forces which can explain the emergence of industrial clustering, namely: geographical and industrial spillovers. In particular we modelled geographical spillovers as the effects caused by the industrial specialisation of a larger area (i.e. a SLA for a FLA), and industrial spillovers as the effects caused by the specialisation of the area in other high-tech sectors. We also tried to take into account the role played by non linearities in agglomeration and scale economies, to explain why firms and cluster do not grow infinitely.

The variables which have been used (with few algebraic transformations) in the regressions are the following:

Dependent variable:

LQM_{iF} = index of industrial specialisation, calculated as the employment location quotient[11] for industry *i* in FLA *F*). The employment location quotient has been chosen between various candidates, as the index for industrial specialisation, because it takes into account the size differences which exist among different establishments, different FLAs and the relative industrial structure of different countries.

Independent variables:

SC_{iF} = index of economies of scale (calculated as the ratio between the employment and the number of establishments in industry *i* and in FLA *F*). The cho-

[10] Which can be linearly estimated through a simple double-log transformation.

[11] As defined in expression (2.4) (see section 2.2.2).

sen variable is therefore the industry and region specific average size of establishment.

AG_{iF} = index of economies of agglomeration (calculated as the ratio between the number of establishments in industry i and in FLA F and the number of manufacturing establishment in the country). The chosen variable weights the number of industry- and region-specific establishments with the number of manufacturing establishment in a given country.

$RLQM_{iS}$ = index of geographical spillovers (calculated as the employment location quotient for industry i in the higher geographical level SLA S, with the exclusion of the FLA under analysis). The chosen variable, in principle, would allow one to identify both positive and negative effects (locational shadowing) which derive from the industrial specialisation of the surrounding areas[12].

$LQMHT_{iF}$ = inter-industry-linkages (calculated as the employment location quotient for the other high-tech sectors — i.e. excluding industry i — in FLA F). This variable measures the extent of inter-industry spillovers which are assumed to be more relevant within the group of high-tech industries.

For each industry i, the estimated model is therefore multiplicative and becomes additive in the double logarithmic specification:

$$LLQM_{iF} = a + bLSC_{iF} + cLAG_{iF} + dLRLQM_{iS} + eLLQMHT_{iF} + \eta_{iF} \qquad (6.4)$$

where all variables are in logs and η_{iF} is a white noise error term, assumed well behaved.

We further tested for the existence of non linearities (i.e. of scale and agglomeration diseconomies after a certain threshold) by introducing two further variables, these being the square of the scale and of the agglomeration indexes. The best results were obtained in the linear additive specifications, where almost all squared coefficients had the expected negative sign and were significantly different from zero. However these specifications had to be rejected on the base of failed tests for mispecification based on the normality of residuals[13].

6.3.3 The data

In order to estimate the relative importance of scale versus agglomeration economies we used part of the original data-set illustrated in the second chapter to estimate six industries specific cross-section versions of equation 6.2 based on first level areas (FLA) for 4 countries.

For reasons of international comparison we restricted our analysis to 5 sectors: 244 (Pharmaceutical), 300 (Computers and office equipment), 321 (Electronic

[12] A better indicator for such a phenomenon would have been the relative specialisation of the bordering geographical FLAs. However the use of SLA is justified in terms of computational convenience and in terms of the hierarchical decision methods used by managers when taking location decisions (as reported by Premus, 1982).

[13] For these regressions no results will therefore be reported.

components), 330 (Instruments), 353 (Aerospace), plus a macro-sector HT (total high-tech) which is the sum of all previously quoted ones.

The chosen functional form requires all variables to be strictly positive. We thus faced the alternatives of either dropping the observations where variables were equal to zero (introducing a sample bias) or substituting the zeros in the data-set with a small value marking these observations with dependent dummy variables. Although the second alternative seems neater we chose to drop the zero-variables observations because of the very nature of the variables. Each time the LQM_{iF} is equal to zero, SC_{iF} and AG_{iF} are also equal to zero, because when in a FLA there are no establishments in a certain industry, there are no employees too.

Because of the nature of the data[14], and the chosen variables and functional form, we were thus unable to use all 306 observations available at the FLA level in our dataset. Each industry specific cross-section regression has therefore been estimated on a specific subset whose size spanned from a maximum of 255 (for the total high-tech sector) to a minimum of 162 (for the Aerospace industry) observations.

6.3.4 The results

Table 6.5 summarises the result of six industry-specific regressions. In general scale economies seem to prevail over agglomeration economies (the scale parameters are about 5 time larger[15]), the exception being sectors 330 and the HT (where agglomeration parameters are, respectively, 3 and 1.5 times larger)[16].

[14] For a great majority of the UK counties data on sectoral employment have been withheld by the ONS for confidentiality reasons (see chapter 2 for details).

[15] In the Electronic components industry, the agglomeration economies parameters is not significantly different from zero, therefore this ratio cannot be calculated.

[16] The results for HT are heavily influenced by the Instruments industry. This is due to the fact that this industry, which is characterised by a low average firm size, records the largest proportion of the total number of firms in high technology industries (64% of the total data-set).

Table 6.5 Scale versus agglomeration economies in the location of high-tech firms

	244	300#	321#	330	353	HT
	Pharmaceut.	Computers	E'tronics	Instruments	Aerospace	high-tech
constant	-3.86	-4.37	-2.43	-0.45	-4.18	-0.76
t-ratio	-1639.00	-25.57	-11.83	-3.81	-11.50	-6.14
LSCiF	0.95	0.82	0.51	0.11	0.81	0.17
t-ratio	22.12	17.94	10.88	3.63	14.56	5.21
LAGiF	0.20	0.13	0.03	0.34	0.17	0.27
t-ratio	5.06	4.01	1.07	9.54	3.17	7.41
LLQMHTiF	0.30	0.49	0.56	0.35	0.53	-
t-ratio	5.21	6.03	8.77	5.83	5.16	-
LRLQMiS	0.21	0.17	0.36	-0.10	0.16	0.56
t-ratio	3.95	3.43	5.20	-2.11	1.89	7.76
Adj. R sq.	0.85	0.79	0.73	0.58	0.77	0.52
n. of obs.	214	204	239	247	162	255

# indicates that regression parameters have been estimated using White's corrected estimator, signal that industrial spillovers could not be calculated for total high-tech. All coefficients are significant at the 5% l.o.s. ($LRQM_{if}$ for Aerospace (353) is significant at 10%).

R^2 for the specific sectors are always very high (spanning from 0.73 to 0.85) while for sectors 330 and HT the values are lower (between 0.52 and 0.58) although still significantly high for cross-section analyses. All regression equations have been tested for functional form mispecification (Ramsey RESET test), non normality of residuals (skewness and kurtosis of residuals), heteroskedasticity (Breusch and Pagan test), and multicollinearity (correlation ratios between dependent variables and auxiliary regressions) and no deviation from the classical model could be detected, apart from a slight heteroskedasticity for industries 300 and 321 which has been corrected by using White's procedure.

It is interesting to note that scale economies seem to perform a larger role in Pharmaceuticals (this is consistent with the structure of the industry where larger plants prevail), while agglomeration economies largely contribute to the geographical concentration of the Instrument industry (where customised production and tailor made products are the rule)[17]. In absolute terms, scale economies are very relevant also for Computers and office machinery and Aerospace. In relative terms, for the Electronic components industry, while scale economies effects are not as important as in other high-tech sectors, agglomeration economies play an absolutely insignificant role in determining the industrial specialisation of the area. However when the complex of all high-tech sectors is being considered, then the agglomeration economies seem, on average, to be more important than scale economies.

By using a double log specification, regression parameters can be interpreted as elasticities. In the Pharmaceutical industry, therefore, a 1% increase of the average establishment size causes an 0.95% increase in the industry location quotient,

[17] One must also consider that Instruments industry (especially in Italy and France) includes many different sub-sectors whose technological level (in certain countries) is not very high.

while the same amount of increase in the relative number of establishments raise the industry's location quotient by a mere 0.2%. It is also worth noting that none of the parameters is higher than the unity.

A further remark concerns the relevance of inter-industry linkages. All regression equations record a positive coefficient for this variable even if this phenomenon appears to be stronger in the Electronic components and in the Aerospace industries, which are benefiting more from the closeness to other high-tech sectors. One can explain this result in term of strong forward linkages for Electronic components (whose products are inputs in several high-tech industries) and of backward linkages for Aerospace (which uses, as inputs, several products of other high-tech sectors).

The final test concerns the presence of spatial positive spillovers or the emergence of locational shadowing. For all but one sector $LRLQM_{iS}$ displays a positive and significant coefficient signalling that a positive relation exists between the FLA specialisation in an industry and the specialisation of the corresponding SLA. Instruments industry displays the existence of locational shadowing phenomena, while and Aerospace records such a low (even though positive) value of the geographical spillovers coefficients and of the relative t-ratio, that one would infer the insignificance of any role played by geographical spillovers, both in positive and negative terms.

6.4 The growth of an isolated cluster

If the set of potential entrants for one specific cluster were independent and separated (had no overlap) from the sets of potential entrants relative to other clusters, then the easiest estimation procedure for analysing the development path of a cluster would imply the estimation of equation (4.5a) which, for convenience, we report here as (6.5)

$$\frac{dn_q(t)}{dt} = r_q n_q(t)\left(1 - \frac{n_q(t)}{K_q}\right) \qquad (6.5)$$

however, such formulation obliges one to determine an exogenous value for the maximum dimension of the cluster (K_q).

An alternative formulation of 4.7b, which has been used in section 4.2.4 for dealing with bi-regional interactions, allows one to overcome such limitations. In this way it is possible to express the rate of variation of the cluster industrial mass as a function of the number of incumbents without any exogenously determined parameter.

$$\frac{dn_q(t)}{dt} = an_q(t) + b\left(n_q(t)\right)^2 \qquad (6.6)$$

Referring to equation (6.5), $a = r_q$, the incipient rate of growth and $\frac{a}{b} = K_q$ the maximum dimension of the cluster.

One could then estimate an empirical version of equation (6.6) as follows

$$\frac{dn_q(t)}{dt} = an_q(t) + b(n_q(t))^2 + e_q(t) \tag{6.7}$$

where $e_q(t)$ is a white noise error term, assumed well behaved, to obtain the estimated coefficients for K_q and r_q.

Once these estimated coefficients have been obtained, one would use them in a regression where the incipient rate of growth and the maximum dimension of the population are functions of a series of locational factors.

In particular, by using the variables' description of section 6.2, one would estimate the following regression equations for each industrial cluster q (i.e. for the industry i in State s)

$$\begin{aligned} K_{is} &= a + bPOP_{is} + cLAND_{is} + dUNEMP_{is} + eMALEPT_{is} + fMANUF_{is} + gTRD_{is} + \\ &+ hPAT_{is} + iHWAY_{is} + lECOL_{is} + mINCO_{is} + nEXPO_{is} + pTAX_{is} + \varepsilon_{is} \end{aligned} \tag{6.8}$$

where the variables used as regressors in (6.8) measure the endowment of input and infrastructures, which are crucial to the development of high-tech businesses and ε_{is} is a white noise error term, assumed well behaved.

$$\begin{aligned} r_{is} &= a + b\frac{HT_{is}}{MANUF_{is}} + cMETRO_{is} + dUNION_{is} + eECOL_{is} + fFAIL_{is} + \\ &+ gTRD_{is} + nEXPO_{is} + pTAX_{is} + \varepsilon_{is} \end{aligned} \tag{6.9}$$

where the variables used as regressors in (6.9) measure the ability to attract outsiders and generate new high-tech firms and ε_{is} is a white noise error term, assumed well behaved.

However, since we believe that the development of a high-tech cluster is heavily dependent on the external conditions (and in particular on what is happening in other industries within the same region and in other regions within the same industry), we are convinced that such models would give biased and inefficient estimates. A possible way round this problem is the estimation of interaction models which can take into account inter-industries and inter-region dynamics.

6.5 Competition and co-operation in the dynamic of clusters development

According to this approach, the development path of a cluster, which is determined by firms' location decision, can be entirely represented by the interactions which exist between entrants and incumbents in different regions and industries. All pure locational factors are assumed to be either mobile or substituted by the developing industries which, in the words of Storper and Walker (1989) are capable of "producing regions".

6.5.1 A model

The analytical formulation of these interactions when the number of interacting entities is equal to 2 is as follows:

$$\begin{cases} \dfrac{dn_r}{dt} = (a_r - a_{rr} n_r - a_{rs} n_s) n_r \\ \dfrac{dn_s}{dt} = (a_s - a_{ss} n_s - a_{sr} n_r) n_s \end{cases} \tag{6.10}$$

where n_r and n_s are the "economic masses" (number of incumbents) in region r and s and all parameters are as in equation (4.7b). In particular a_r and a_s are the intrinsic rates of increase of each region in isolation; a_{rr} and a_{ss} are the intra-regional competition parameters which reflect the inhibiting effects that a firm's entry has on the growth rate of the same region (because of congestion effects); a_{rs} and a_{sr} (inter-regional competition parameters) show the inhibiting effects that one firm, locating in a region, has on the growth of the other region (inter-regional competition parameters).

One then would like also to take into account the inter-industry interactions which develop within the same region. If this is the case, then a two-regions, two sectors can be modelled as follows

$$\begin{cases} \dfrac{dn_r^i}{dt} = \left(a_r^i - a_{rr}^i n_r^i - a_{rs}^i n_s^i + a_r^j n_r^j\right) n_r^i \\ \dfrac{dn_s^i}{dt} = \left(a_s^i - a_{ss}^i n_s^i - a_{sr}^i n_r^i + a_s^j n_s^j\right) n_s^i \\ \dfrac{dn_r^j}{dt} = \left(a_r^j - a_{rr}^j n_r^j - a_{rs}^j n_s^j + a_r^i n_r^i\right) n_r^j \\ \dfrac{dn_s^j}{dt} = \left(a_s^j - a_{ss}^j n_s^j - a_{sr}^j n_r^j + a_s^i n_s^i\right) n_s^j \end{cases} \tag{6.11}$$

where the subscripts r, s refer to regions, and the superscripts i, j refer to industries. It appears evident that, even though this formulation is linear, and expresses the agglomeration diseconomies without exogenously determined parameters, its applicability is limited by the large number of parameters which have to be estimated. We thus proceeded by regressing the growth of each cluster (i.e. for the couplet industry-region) on the different interaction coefficients (these being both inter-regional and inter-sectoral). The results of these regressions were disappointing: in many cases the inner dynamics of the cluster growth (which are represented by the first two dependent variables in equation 6.11) explained a large part of the variance of $\dfrac{dn_r^i}{dt}$, while the other estimated interaction coefficients were almost always insignificant.

Table 6.6 reports the only valuable results which we were able to extract from a set of over 50 regressions and refers only to inter-state intra-industry relations.

Table 6.6 US high-tech cluster development and inter-state competition coefficients

cluster growth	period of time	cluster size	squared cluster size	first competitor	second competitor	third competitor	fourth competitor	Adj. Rsq
CAL357	1957-1994			TEX357				0.66
		0.254*	-0.00016*	-0.81*				
t-ratio		*3.66*	*-288*	*-204*				
CAL367	1957-1994			PEN367				
		0.19*	-0.12	-1.03*	MAS367	TEX367	ILL367	0.40
t-ratio		*3.78*	*-0.4*	*-349*	*-0.85**	*-0.82***	*-0.52***	
CAL380	1949-1994			MAS380	-3.02	-1.90	-1.96	
		0.18**	0.00036*	-0.31***				0.72
t-ratio		*1.96*	*-258*	*-180*				
MAS357	1957-1994			TEX357				
		0.21*	-0.00016	-0.24*				0.55
t-ratio		*4.3*	*-0.47*	*-287*				
MAS380	1949-1994			PEN380				
		0.17*	-0.0001*	-0.11*				0.59
		253	-202	-218				

* = 5%; ** = 10%; *** = 25% l.o.s.

From table 6.6 it appears that, as far as the Computer industry (357) is concerned, the first three States play very different roles: California is still the leading State but is beginning to suffer from the competition of Texas. Massachusetts, the old leader, is fighting directly with Texas and indirectly with California, while Texas Computer industry grows without any significant interference from other States.

In the Electronic components industry (367) the leadership of California is clear in absolute term, but in the last ten years, when California has incurred some difficulties, other states (such as Pennsylvania, Massachusetts, Texas and Illinois) have shown their competitive strength.

In the Instruments sector (380) — where New York has steadily reduced its presence and California has taken the lead — our estimates signal the existence of a competitive relation between two late-comer States (Massachussets and Pennsylvania), which are progressively catching-up.

As already said, the main difficulty with such an empirical analysis is represented by the exponential increase in the number of regressors to be estimated. The interaction variables — which refer to the number of firms in each other high-tech sector located in the same State and to the number of firms belonging to the same sector located in other States — may easily overcome the number of available observations.

We found a solution to solve this problem by creating two composite variables which, for each high-tech cluster (defined as the couplet of state s and a sector i), summarise the inter-state/intra-industry relations $n_{(S-s)i}$ [18] (i.e. the number of incumbents in all but state s belonging to industry i) and the intra-state/inter-industry relations $n_{s(I-i)}$ (i.e. the number of incumbents in state s belonging to all but industry i). For each cluster we therefore estimated the following regression:

$$\frac{dn_{si}}{dt} = \alpha + \beta_1 n_{si} + \beta_2 n_{si}^2 + \beta_3 n_{(S-s)i} + \beta_4 n_{s(I-i)} + \varepsilon_{si} \qquad (6.12)$$

All regression equations have been tested for serial correlation (Lagrange multiplier), functional form mispecification (Ramsey RESET), non normality of residuals (skewness and kurtosis), heteroskedasticity (Breusch and Pagan), and multicollinearity (correlation ratios between dependent variables and auxiliary regressions) and no significant deviation from the classical model could be detected. Table 6.7 presents the results.

[18] Table 6.7 shows two different measurements of the inter-state/intra-industry relations. The first (absolute) measures the number of incumbents in other significant industry-specific clusters. The second (relative) measures the ratio between the above and the number of US establishments belonging to that particular high-tech industry.

Table 6.7 The development of clusters: inter-regional and inter-industries effects

cluster growth	period of time number of obs.	cluster size	squared cluster square	interstate intraindustry relations	interstate relative relations	intrastate interindustry relations	**Adj. Rsq**
DCAL283	1957-1994	0.40***	0.0004	149.21*		0.14*	**0.37#**
t-ratio	*38 obs.*	*-1.16*	*0.49*	*2.24*		*3.51*	
DCAL283					161.31*		
t-ratio					*2.58*		
DCAL357	1957-1994	0.24***	-0.0002***	-0.13***		0.048	**0.44#**
t-ratio	*38 obs.*	*1.26*	*-1.64*	*-1.36*		*0.32*	
DCAL357					45.54		
t-ratio					*0.15*		
DCAL367	1959-1994	0.32	-0.0002**	-0.04		0.004	**0.22#**
t-ratio	*35 obs.*	*0.64*	*-1.92*	*-0.10*		*0.3*	
DCAL367					-1580.6***		
t-ratio					*-1.19*		
DCAL372	1957-1994	0.02	-0.0001	-0.09***		0.01*	**0.43#**
t-ratio	*38 obs.*	*0.065*	*-0.34*	*-1.60*		*2.24*	
DCAL372					254.69		
t-ratio					*0.63*		
DCAL380	1948-1994	-0.02	-0.00002	0.01		0.04	**0.68#**
t-ratio	*45 obs.*	*-0.12*	*-0.30*	*0.07*		*0.88*	
DCAL380					-527.17		
t-ratio					*0.57*		
DCAL737	1974-1994	0.08	0.00004***	-0.20		0.46	**0.64**
t-ratio	*20 obs.*	*0.74*	*1.33*	*-0.73*		*0.81*	
DCAL737					-0.24		
t-ratio					*-1.51*		
DTEX283	1957-1994	-0.81**	0.0002	0.029*		0.02*	**0.55#**
t-ratio	*38 obs.*	*-1.86*	*0.05*	*3.14*		*3.79*	
DTEX283					172.49*		
t-ratio					*4.26*		
DTEX357	1957-1994	-0.27	-0.00003	0.011		0.03**	**0.34#**
t-ratio	*38 obs.*	*-1.2*	*-0.033*	*0.97*		*1.81*	
DTEX357							
t-ratio							
DTEX367	1959-1994	-0.90*	-0.00005	0.049**		0.27*	**0.53**
t-ratio	*35 obs.*	*-2.68*	*-0.16*	*1.77*		*4.77*	
DTEX367							
t-ratio							
DTEX372	1957-1994	-0.44*	0.0001	-0.14***		0.20	**0.52#**
t-ratio	*38 obs.*	*-2.61*	*0.61*	*1.16*		*0.91*	
DTEX372					-81.09		
t-ratio					*-0.62*		
DTEX376	1974-1994	0.78*	-0.11*	-0.14		0.004*	**0.51**
t-ratio	*20 obs.*	*2.63*	*-3.33*	*-0.33*		*3.76*	
DTEX376					19.29*		
t-ratio					*2.33*		
DTEX380	1948-1994	-0.17	-0.0001	-0.002		0.097***	**0.33#**
t-ratio	*45 obs.*	*-0.48*	*-0.84*	*-0.08*		*1.26*	
DTEX380					-471.75***		
t-ratio					*-1.47*		
DTEX737	1974-1994	0.46	0.00008***	-0.17***		1.36	**0.46#**
t-ratio	*20 obs.*	*0.54*	*1.32*	*-1.43*		*1.12*	
DTEX737					-0.11*		
t-ratio					*-2.96*		
DMAS283	1957-1994	-0.35	0.0003	0.036**		0.012*	**0.44#**
t-ratio	*38 obs.*	*-0.65*	*0.047*	*1.73*		*3.24*	
DMAS283					-14.94		
t-ratio					*-0.32*		
DMAS357	1957-1994	0.35*	-0.002*	0.09*		0.053*	**0.49#**
t-ratio	*38 obs.*	*2.08*	*-2.72*	*2.12*		*2.09*	
DMAS357							
t-ratio							

cluster growth	period of time number of obs.	cluster size	squared cluster size	interstate intraindustry relations	interstate relative relations	intrastate interindustry relations	**Adj. Rsq**
DMAS367	1959-1994	-0.16	-0.0004	-0.00009		0.18**	**0.05**
t-ratio	*35 obs.*	*-0.54*	*-0.89*	*0.0023*		*1.54*	
DMAS367					-309.05		
t-ratio					*0.59*		
DMAS376	1974-1994	-0.33	0.04	-0.021		0.002*	**0.57#**
t-ratio	*20 obs.*	*-0.65*	*0.51*	*-0.56*		*2.25*	
DMAS376							
t-ratio							
DMAS380	1948-1994	-0.54	-0.0003**	-0.029		0.033	**0.38#**
t-ratio	*45 obs.*	*-0.26*	*-1.95*	*-1.12*		*0.82*	
DMAS380					-316.68***		
t-ratio					*-1.16*		
DNYO283	1957-1994	-0.73*	0.0019*	-0.058*		0.039	**0.33**
t-ratio	*38 obs.*	*-2.69*	*2.47*	*-2.13*		*2.23*	
DNYO283					-123.92***		
t-ratio					*-1.31*		
DNYO737	1974-1994	-1.02**	-0.00008	0.26*		-0.25	**0.36**
t-ratio	*20 obs.*	*-1.98*	*-1.18*	*2.86*		*-0.21*	
DMAS737					-0.36		
t-ratio					*-0.81*		
DFLO372	1957-1994	0.31**	-0.0069*	0.023*	82.30**	0.11	**0.48**
t-ratio	*38 obs.*	*1.32*	*-4.29*	*3.14*	*1.90*	*5.06*	
DFLO372							
t-ratio							
DFLO380	1948-1994	0.12	-0.0002*	-0.0026		0.27***	**0.66#**
t-ratio	*45 obs.*	*1.07*	*-2.59*	*-0.25*		*1.19*	
DFLO380					2.52		
t-ratio					*0.024*		
DCOL357	1957-1994	-0.31	-0.0008	0.016**			**0.32#**
t-ratio	*38 obs.*	*-1.04*	*-0.32*	*1.97*			
DCOL357					42.23		
t-ratio					*0.82*		
DILL737	1974-1994	1.80*	-0.00014*	-0.11*		-2.15**	**0.68#**
t-ratio	*20 obs.*	*3.00*	*-2.68*	*-2.05*		*-1.44*	
DILL737					-0.0055		
t-ratio					*-0.15*		
DWAS372	1957-1994	0.21***	-0.0013	0.0044			**0.56#**
t-ratio	*38 obs.*	*1.21*	*-0.99*	*0.52*			
DWAS372					19.86		
t-ratio					*0.40*		
DNJE 737	1974-1994	0.80***	0.0001*	0.21*		0.95**	**0.91#**
t-ratio	*20 obs.*	*1.58*	*3.92*	*2.63*		*1.80*	
DNJE 737					-1474***		
t-ratio					*-1.72*		
DPEN367	1959-1994	-0.28	0.0002	0.015*		0.01	**0.47**
t-ratio	*35 obs.*	*-1.35*	*0.49*	*2.43*		*0.31*	
DPEN367					195.62***		
t-ratio					*1.39*		
DUTA376	1974-1994	0.47	-0.09***	0.059*			**0.54#**
t-ratio	*20 obs.*	*0.73*	*-1.20*	*2.92*			
DUTA376					-0.92		
t-ratio					*-0.15*		

* = 5%; ** = 10%; *** = 25% l.o.s.; # regressors include also one or two dummies for particular years

The results show that at 5% l.o.s. the extent of inter-cluster competition (i.e. different States competing as alternative locations for the same pool of potential industry-specific entrants) is limited. Significant negative coefficients are displayed only by 3 clusters: Computer service and data processing industry (737) in Texas and Illinois and Drugs (283) in New York. However, with a larger l.o.s. (equal to 25%), the number of significant negative coefficients are recorded also by three industries in California (Computers (357), Electronic components (367) and Aircraft (372)), two industries in Texas (Aircraft and Instruments (380)), Instruments in Massachusetts, and Computer service and data processing in New Jersey.

On the contrary, the results show a stronger influence of intra-cluster synergies (i.e. the level of other high-tech industries in the same state positively influence the development of a particular cluster). At the 5% l.o.s., 8 out of 28 intra-cluster regressors[19] show significant and positive coefficient, while only Computer service and data processing industry in Illinois display a negative coefficient which may be interpreted as a sign of Inter-industry competition on some fixed pool of generic resources (capital and/or real estates).

The empirical exercise shows also that, when interaction coefficients are taken into account, very few regressions[20] display the expected signs for both the number of incumbents (cluster size) and its squared value (squared cluster size).

6.6. Market structure and industrial specialization in the growth of high-tech clusters

This section focuses on the empirical investigation of the claim that externalities — and in particular technological externalities, which are associated with knowledge spillovers — are the "engine of growth" of high-tech clusters. In particular, following Glaeser et al. (1992), we will test the empirical relevance of three alternative explanations which draw their respective theoretical background from three different streams of literature, these being economic growth, strategic management, and economic history.

The first approach stresses the role of what Glaeser et al. (1992) call "Marshall-Arrow-Romer (MAR) externality". According to this view "concentration in a city-industry helps knowledge spillovers between firms and therefore the growth of that industry and of that city" (Glaeser et al. 1992, p. 1127). This approach also predicts that "local monopoly is better for growth than local competition because it allows externalities to be internalised by innovators. When externalities are internalised, innovation and growth speed up" (ibid.)[21]. Therefore the AR story reads

[19] Which become 13 at 25% l.o.s.

[20] Five clusters: Computers in California and Massachusetts, Missiles in Texas, Aircraft in Florida, and Computer services in Illinois.

[21] This last statement, surely in line with Arrow (1962) and Romer (1986), is, in my opinion, far away from what Marshall (1921) thought about the inner dynamics of growth. For this reason, in the following section we will refer to this approach as Arrow-Romer (AR) approach. Incidentally one may also note that the AR's argument is similar to most of Krugman's ideas on this issue.

as follows: geographical specialisation and monopolistic power are the best conditions for knowledge externalities and, therefore, for city (or cluster) growth.

The second approach refers to the contribute of Porter (1990). According to Porter knowledge spillovers within geographically concentrated industries stimulate growth. However he also insists that "local competition as opposed to local monopoly foster the pursuits and rapid adoption of innovation" (Glaeser et al. 1992, p. 1128)[22]. The role of internal competition is crucial in Porter's diamond of competitive advantages[23], in order to prevent technologically advanced industry from resting on past successes and becoming obsolete. Porter's story is the following: geographical specialisation and fierce competition are the best conditions for knowledge externalities and, therefore, for cities (or clusters) growth[24].

The third approach derives from two contributes of Jacobs (1969 and 1984) which underline that, in general, "the most important knowledge transfers come from outside the core industry" (Glaeser et al. 1992, p. 1128). Jacobs logically shows, and gives many historical examples, that variety and diversity of geographically proximate industries[25] together with local competition favours growth. Thus, for Jacobs, local differentiation of the industry and fierce competition are the best conditions for knowledge externalities and, therefore, for city (or cluster) growth[26].

To test these hypotheses, Glaeser et al. (1992) build a specific data-set on the 6 largest industries (2 digit SIC) in the top 170 standard metropolitan areas of USA and test a model which explains the growth rates of employment and wage (for the period 1956-1987) in a sample of 1016 city-industry, in terms of various combination of the three above mentioned main dimensions: geographical specialisation vs. geographical diffusion, industrial differentiation vs. industrial concentration, and competition vs. monopoly. We found this article very interesting and, following Baptista and Swann (1996) which perform a similar analysis on US computer firms' clusters, we decided to test a modified version of the model which better suited the data and the specific issue of study in the thesis: the development of high-tech clusters. For this reason some of the original variables have been modified in order to take into account specific definitions of specialisation, and competition within the high-tech sectors.

---

[22] "Rivalry among firms with the same home base is particularly beneficial for a variety of reasons. First, strong domestic competition create particular visible pressure on firms to improve. It also often attracts new rivals to the industry. (...) Geographical concentration of rivals in a single city or region within a nation both reflects and magnifies these benefits" (Porter 1990, p.119-120).

[23] See section 3.8.

[24] Porter thus highlights the role played by localisation externalities.

[25] "The great capitals of modern Europe did not become great cities because they were capitals. Cause and effects ran the other way. (..) Paris, Berlin and London became the genuine capitals only after they had already become the largest (and economically the most diversified) commercial and industrial city of their Kingdoms" (Jacobs 1969, p. 143).

[26] Jacobs thus highlights the role played by urbanisation/regionalisation externalities.

6.6.1 A model

The three above mentioned theories can be formalised in a simple economic model that will be then empirically tested in order to discriminate between the conflicting explanations. Following Glaeser et al. (1992), suppose that a firm belonging to a given industry in a certain location has a production function given by $pY = A_t f(l_t)$ where p is a price index, Y is real output, A represents the overall level of technology at time t (which reflects changes in both technology and price) and l_t is the labour input at time t[27]. Each firm in the industry takes technology, prices and wages w_t as given, and maximises:

$$A_t f(l_t) - w_t l_t \tag{6.13}$$

the first order conditions are therefore

$$A_t f'(l_t) = w_t \tag{6.14}$$

which can be rewritten in terms of growth rates as

$$\log\left(\frac{A_{t+1}}{A_t}\right) = \log\left(\frac{w_{t+1}}{w_t}\right) - \log\left(\frac{f'(l_{t+1})}{f'(l_t)}\right) \tag{6.15}$$

The level of technology in the region-industry is assumed to have both local and national components as follows:

$$A = A_{\text{local}} A_{\text{national}} \tag{6.16}$$

The growth rate will then be the sum of the growth of national technology in the industry and the growth of local technology

$$\log\left(\frac{A_{t+1}}{A_t}\right) = \log\left(\frac{A_{\text{local},t+1}}{A_{\text{local},t}}\right) + \log\left(\frac{A_{\text{national},t+1}}{A_{\text{national},t}}\right) \tag{6.17}$$

The growth in national technology is assumed to capture changes in the price of the product and shifts in the national technology for that industry, while the local technology is assumed to grow at a rate which is exogenous to the single firm but depends on the different technological externalities which are present in the region-industry.

$$\log\left(\frac{A_{\text{local},t+1}}{A_{\text{local},t}}\right) = g(\text{specializ., local monop., diversif., initial condit.}) + e_{t+1} \tag{6.18}$$

If we set $f(l) = l^{1-\alpha}$, where $0 < \alpha < 1$, and we combine (6.15), (6.17), and (6.18) we obtain:

[27] By choosing such functional form for the production function, which abstracts from capital inputs, "we may not capture labour-saving technological innovations and we shall not capture innovations that result only in further accumulation of physical capital" (Glaeser et al. 1992, p. 1132).

$$\log\left(\frac{l_{t+1}}{l_t}\right) = -\log\left(\frac{w_{t+1}}{w_t}\right) + \log\left(\frac{\mathrm{A}_{\mathrm{national},t+1}}{\mathrm{A}_{\mathrm{national},t}}\right) + g\,(s,\, lm,\, d,\, ic) + \varepsilon_{t+1} \qquad (6.19)$$

where s stands for specialisation; lm for local monopoly; d for diversification, and ic for initial conditions and ε is a white noise error term, assumed well behaved.

Growth in nation-wide industry employment is assumed to capture changes in nation-wide technology and prices; workers are assumed to participate in a nation-wide labour market[28] so that wage growth will be a constant across state-industries. Equation (6.19) then allows one to associate the growth of employment in a state industry (or cluster q) with measures of the technological externalities proposed by the different theories as follows:

$$\begin{aligned} &LGMSI_q = aLGMUI + bLWIN75_q + cLMIN75_q + dSPEIN_q + \\ &+ eCOMINHT_q + fDIVINHT_q + \varepsilon_q \end{aligned} \qquad (6.20)$$

where $SPEIN = SPEINHT$ in model 1 and $SPEIN = SPEINTOT$ in model 2, 5 and 6 (see table 6.9).

The variables of the estimated version of model has been based on the following variables.

Dependent variable:

$LGMSI_q$ = Log (Employment in 1995/ employment in 1975) in the state-industry.

Independent variables:

$LGMUI$ = Log (US employment in 1995/US employment in 1975) in the industry.

$LWIN75_q$ = Log of wage in the state-industry in 1975 in thousand dollars per years (calculated as the ratio between the annual payroll and the employment in the state -industry).

$MIN75_q$ = Log of employment in the state-industry in 1975.

$SPEINHT_q$ = index of state-industry relative geographical specialisation within the HT macrosector in 1975 (calculated as a modified employment location quotient were the industrial total is calculated only on the high-tech part)[29].

$SPEINTOT_q$ = alternative index of state-industry geographical specialisation in 1975 (calculated as a generic location quotient based on employment).

---

[28] This assumption is crucially dependent on the US institutional framework. The empirical estimates will thus be produced on US data.

[29] SPEINHT = (employment in industry i in state s / total high-tech employment in state s) / (US employment in industry i / total high-tech employment).

$COMINHT_q$ = index of relative local intrasectoral competition in 1975 (calculated as the ratio of the number of firms per workers in industry *i* in state *s* and the number of workers in the same industry in the US)[30].

$DIVINHT_q$ = index of relative diversification of the state within the high-tech macrosector in 1975 (calculated as a modified Herfindhal index where the shares are calculated on the remaining high-tech sectors). Because of its formulation $DIVINHT_q$ is an inverse index of diversification. The lower is the value of the index, the higher is the level of diversification of the state within the group we classified as high-tech industries.

6.6.2 The data

The data used in this exercise is the original time series dataset we built from County Business Patterns data in order to test the macro-ecologic approach. Since we were able to collect historical data (at this fine level of spatial and sectoral level of disaggregation) only for the US, we did not have to worry about international comparisons but only about the changes in the US industrial classification definition, SIC (the most recent occurring in 1988).

The final dataset is composed of 112 observations on the growth rate of different variables between 1975 and 1995 for 7 sectors within 16 US States[31]. The sectors involved in the analysis are (in the original US SIC codes and definitions): 283, Drugs; 356, Office and computing machines; 367, Electronic components and accessories; 372, Aircraft and parts[32], 737 Computer and data processing services, 7391 Research and development laboratories[33].

---

[30] COMINT = (firms in industry *i* in state *s*/employment in industry *i* in state *s*) / (firms in industry *i* in the US / employment in industry *i* in the US).

[31] These 16 states have been selected as the most technologically advanced in the US (i.e. the one with the largest number of employees or establishment in the high-tech sectors) either in 1956 or in 1994. The complete list is as follows: Arizona, California, Colorado, Connecticut, Florida, Illinois, Kansas, Massachusetts, Michigan, New Jersey, New York, Ohio, Pennsylvania, Texas, Utah, Washington. From a geographical perspective they offer a non-biased sample of both the 9 Census divisions and of the 4 Census regions. For withheld employment data we used the class size average, for each payroll missing data we constructed an "artificial" payroll data by using the employment figures obtained as above and a wage proxy calculated on "similar" sectors. (i.e. when employment and payroll data for Colorado in 1975 were missing for industry 372 Aircrafts and parts, we used the average of the employment size class and an artificial payroll figure calculated as the product of the "average" employment and the wage of sector 3724 Aircrafts engine and engine parts).

[32] Sector 376, Guided missiles, space vehicles, and parts, had to be dropped because most of the figures on employment were withheld for privacy reasons.

[33] This sectors became 8371 Commercial physics research, after the changes in SIC occurred in 1988. We are grateful to Y.D. Funderburk (US Census Bureau) for helping us in matching precisely the sectors through the different SIC codes and for supplying some additional 1975 data that we did not collect in our original data-set.

Table 6.8 shows the four fastest and slowest growing state-high-tech industries in the period 1975-95.

Table 6.8 Fastest and slowest growing state-industries (employment) in the period 1975-1995

States	*Industries*	*lgwsi*	*speinht*	*speintot*	*cominht*	*divinht*
COLORADO	Computer services	2.714	1.022	1.282	1.399	0.360
UTAH	Instruments	2.702	0.399	0.307	4.496	0.277
UTAH	Computer services	2.598	1.710	1.316	1.167	0.292
MASSACHUSETTS	Computer services	2.585	0.575	0.881	1.878	0.294
average fastest growing industries		**2.650**	**0.927**	**0.946**	**2.235**	**0.306**
KANSAS	Computer and o.m.	-1.285	0.352	0.790	0.692	0.725
NEW JERSEY	Computer and o.m.	-1.290	0.633	0.822	1.873	0.262
NEW YORK	Aircrafts	-1.560	0.550	0.779	0.910	0.290
OHIO	Computer and o.m.	-1.595	0.891	0.693	0.701	0.305
average slowest growing industries		**-1.433**	**0.607**	**0.771**	**1.044**	**0.395**
differences		***4.082***	***0.320***	***0.176***	***1.191***	***-0.090***

Table 6.8 gives three impressions. First, rapidly growing state-industries were more geographically concentrated (both in relative[34] and in absolute terms) than rapidly declining ones. Second fast growing state-industries were more competitive than shirking state-industries. Third, the effects of diversification on the growth performance of state-industries is positive (i.e. more diversified clusters grow more than the others) but weak[35].

The table seems therefore to support the interpretation on the genesis of externalities, proposed by Porter, which states that the engines of growth are geographical specialisation and local competition. However, a detailed econometric analysis is needed in order to test, in greater detail, the above mentioned hypotheses.

[34] With respect to the high-tech industry as a whole.

[35] The table also shows the westward shift of high-tech industries (fastest growing state-industries are in the Mountains Census Division) and the different stages of development of industrial high-tech industries (with computer services rapidly growing and computer manufacturing declining). This last result can be seen also as a prove of the tertiarisation of advanced economies.

6.6.3 The results

Table 6.9 shows the results for different empirical specifications of the model. R^2 coefficients of every model are quite high for a cross-section analysis (they vary between 0.65 and 0.69) and stable. The best fit of the regression is achieved by model 6 (which, as will be discussed below, test as the same time both the AR's and the Porter's theories).

All regression equations have been tested for serial correlation (Lagrange multiplier test), functional form mispecification (Ramsey RESET test), non normality of residuals (skewness and kurtosis of residuals), heteroskedasticity (Breusch and Pagan test), and multicollinearity (correlation ratios between dependent variables and auxiliary regressions) and no significant deviation from the classical model could be detected, apart from heteroskedasticity which has been corrected by using White's procedure.

Table 6.9 The engines of high-tech clusters' growth

models	1*	2*	3*	4*	5*	6*	7*
constant	4.34	4.69	2.82	4.71	3.34	2.9	3.19
t-ratio	*4.86*	*5.25*	*2.48*	*5.54*	*3.18*	*2.61*	*2.99*
LGMUI	0.99	0.99	1.05	0.98	1.08	1.08	1.05
t-ratio	*13.16*	*13.51*	*12.94*	*13.3*	*13.75*	*13.55*	*13.08*
LWIN75	-0.84	-0.89	-0.67	-0.85	-0.71	-0.69	-0.68
t-ratio	*-2.35*	*-2.5*	*-1.86*	*-2.5*	*-2.12*	*-1.95*	*-1.98*
LMIN75	-0.25	-0.29	0.16	0.27	-0.22	-0.19	-0.18
t-ratio	*-5.56*	*-6.45*	*-3*	*-6.33*	*-4.07*	*-3.62*	*-3.44*
SPEINHT	0.01						
t-ratio	*-0.07*						
SPEINTOT		0.09			0.15	0.14	
t-ratio		*2.78*			*3.79*	*3.63*	
COMINHT			0.18		0.22	0.22	0.17
t-ratio			*2.07*		*2.53*	*2.55*	*2.02*
DIVINHT				-0.53	-0.58		-0.48
t-ratio				*1.12*	*-1.28*		*-1.07*
Adj. R sq.	0.65	0.66	0.67	0.65	0.69	0.68	0.67
n. of obs.	112	112	112	112	112	112	112

* indicates that the regression parameters have been estimated using White's corrected version of the variance-covariance matrix of parameters. Al regression coefficients are significant at 5% l.o.s. (a part from LGWIN75 in model 6 and DIVINHT in all models which are both significant only at 20% l.o.s.)

The different specifications allow one to identify both the single and the combined effects of the different kinds of externalities on the growth of high-tech clusters. All specifications include a set of control variables such as: national employment growth (*LGMUI*) which corrects for demand shifts, log of wage in the state-industry in 1975 (*LGWIN75*) which corrects for generic movement of firms towards low-wage areas, and log of employment in the state-industry in 1975 (*LGMIN75*) which corrects for generic movement of workers toward high-employment state-industry. Alternatively and somehow more sensibly, one can interpret the negative coefficients on the initial values of employment and wages as an indirect way to measure the geographical westbound shifts of high-tech industries in the US from the old industrial East coast towards the Mountains and the Pacific (where these sectors were still in the first development phases and where wages were lower). All control variables have the expected signs. High initial employment and high wages in a cluster lead to slower growth (negative coefficients for both *LWIN75* and *LMIN75*); while employment growth in the cluster is positively correlated with US wide industry growth.

Table 6.9 presents six different empirical specifications of the same basic model. The first three specifications look at each externality effect in isolation while the remaining ones test the empirical relevance of the three theories through a couple of coefficients.

The first specification uses *SPEINHT* as the index of geographic specialisation, and the parameter is not significantly different from zero. In the second specification we use the alternative index for geographic specialisation *SPEINTOT* (where the state-industry is compared with the total of manufacturing and service activities) and we find the coefficient to be positive and significant. Thus, sectoral specialisation in specific high-tech industries fosters growth. This result is consistent with both AR and Porter and in contrast with Jacobs.

The third specification analyses *COMINHT*, which is an index of competition (or, alternatively, an inverse index of monopoly power). The parameter is positive and significant suggesting a positive correlation between the extent of local competition and the growth of the cluster[36]. This result is thus consistent with Porter and Jacobs but in contrast with AR.

The fourth specification, by introducing *DIVINHT*, studies the effect of industrial diversification. Although the parameter had the correct negative sign (meaning that a lower value of the Herfindhal index, which means a higher diversification in the other h-t sectors, is positively correlated with the cluster growth rate), it shows a lower significance[37]. Thus we have no strong empirical support for an important part of the Jacob's theory which, for high-tech industries, will take count of the inter-industry (i.e. within the whole high-tech sector), technological externalities and knowledge spillovers[38].

The fifth specification tests the significance of all parameters together (*SPEINTOT*, *COMINHT*, and *DIVINHT*). The empirical evidence thus supports the theory put forward by Porter, and here adapted to the case of high-tech industries, which considers intra-industry technological spillovers and intra-industry local competition and rivalry as the spurs to innovation, growth and success of clusters[39].

---

[36] This is consistent with lots of anecdotal literature on the business climate of successful high-tech clusters such as Silicon Valley, Research Triangle etc.

[37] It is always significant only at the 20% l.o.s.

[38] The existence and the relevance of these spillovers has been shown by Scherer (1982) presenting systematic evidence that around 70% of innovation in a given industry are used outside that industry.

[39] This result is therefore in contrast to Glaeser et al. (1992) and consistent with Baptista and Swann (1996).

The last two specifications test the three theories in a more formal way (each time with only two relevant parameters). The estimated coefficients of model 5 give stronger support to Porter's theory since both values of *SPEINTOT* and *COMINHT* are positive and significant; and are definitely against AR. Model 6 is partially supporting Jacobs in the sense that while the coefficient of *COMINHT* is positive and significant, the coefficient of *DIVINHT* has the expected (negative) sign but its statistical significance is low.

6.7 Why do high-tech firms cluster? Some empirical considerations

If one jointly considers the empirical results of section 6.3 and 6.6, then a descriptive conclusion would merely state that, while the actual degree of innovative industrial specialisation of an area (see section 6.3) is heavily dependent on the presence of large establishments (which enjoy higher economies of scale), the growth rate of its high-tech employment (see section 6.6) is fostered by the presence of a competitive environment where small firms are predominant.

A tentative explanation of the different relations existing between the average firm's size (in a given industry and region), the degree of specialisation of a region, and the growth rate of its industry employment reads as follows. The relative specialisation of a given area (as measured by the industry location quotient) is explained in terms of a larger average size of firms. Large high-tech firms are more efficient (because of the existence of economies of scale in some part of the production process, say in R&D activities) and they acquire growing market shares. The region becomes nation-wide acknowledged as the centre of the industry, and the regional level of industry employment increases at the expenses of other industries (causing sectoral changes in the regional industrial structure of production) and/or of other regions (causing regional changes in the national geographic structure of production, and the inflow of skilled high-tech industry workers). The negative effects of scale economies on employment are in fact more than compensated for by the increasing market share. The process continues up to the point where regional firms as a whole have reached a dominant position in the market and the growth of any single firm is obtained at the expenses of the remaining firms in the region. From this point onward any increase in the average firm size necessarily implies a process of regional concentration and therefore, because of economies of scale, the reduction of the employment growth rate[40].

An alternative explanation, which directly relates to the organisation ecology literature (Hannan and Freeman 1989) and that we have adapted to the development of industrial clusters is the following. In the early stages of development of a cluster small and pioneers firms enter from outside the region and/or are generated by a spin-off process from existing firms (which may well be doing business in

[40] Eventually, as the process continues, the regional industry employment growth rate decreases until it become negative and the industry employment level in the region is reduced.

other industrial sectors). Thus the positive relation between growth and small firms size in the early "heroic" stages of development of an industrial cluster is explained. However, when the industry begins to be important in the region's economy and the first signal of input competition and congestion begins to appear, then these small and creative firms stop being the fittest organisational form. They are gradually substituted by larger and more structured firms which, by enjoying scale economies in a technology that is gradually becoming "stable", make full use of their superior organisational and market power to make profits and grow. The locational pattern of established high-tech industries is therefore determined by economies of scale much more than by agglomeration economies.

Finally it is worth to stress that the empirical analyses performed in this chapter have add new contributions to the existing knowledge of the structure and dynamics of innovative industrial clusters in many respects. In particular:

1. The analyses performed in section 6.2 confirmed the positive influence of the existence of a local pool of skilled labour force (Marshall 1920; Krugman 1991a), the strength of the technological infrastructure (Feldman 1995), the degree of trade openness of the cluster (Porter 1996), and a negative influence of corporate income tax rate and unemployment rate (Storper and Walker 1989; Hirshleifer 1993).
2. The analyses of section 6.3, to my knowledge, are the first empirical studies devoted to measure the relative importance of scale versus agglomeration economies for high-tech sectors and they are based on an extremely wide database. Furthermore in the same section an analysis of inter-industry and inter-cluster spillovers has been performed together with an empirical test of the relevance of locational shadowing (Arthur 1990).
3. In the analyses of section 6.5, which bear some similarities to those performed by Swann (1998) and Swann and Prevezer (1996), we attempted to give empirical consistency to the ecological approach (Dendrinos and Mullally 1985; Hannan and Freeman 1989; Nijkamp and Reggiani 1998) to spatial economics and location theory.
4. In the analyses of section 6.6 we applied the original empirical framework proposed by Glaser et al. (1992) to the growth of US high-tech cluster, obtaining some interesting results which support Porter's (1990 and 1996) explanation.

Chapter 7 - Policy implications

> The evolving patterns of uneven regional development are not simply reflections or outcomes of the long waves in the national economy. They are the process of national economic change and development.
>
> M. Marshall (1987), *Long Waves of Regional Development.*

7.1 High-tech firms and local development policies

In recent times, almost all industrialised countries are experiencing, in varying degrees, two contemporary phenomena: the crisis of traditional industries and the development of new innovative sectors. These phenomena, which are caused by national and international determinants and spurred by the dynamics of globalisation and economic interdependence, have important consequences at the local level, the most relevant being the emergence and deepening of regional differentiation (Doz 1987).

The process of innovation and industrial change is following regional paths as are crisis and decline. The same process of geographic concentration of science based industries, which yields welfare in certain areas, is causing crises of the productive structure and increases in unemployment in others. This remark has a crucial importance for most European countries while they experience a process of economic and monetary[1] integration. Whether the formation of EMU will cause more or less agglomeration in the high-tech industries is a question highly debated in the literature. Whether national and European authorities should view the agglomeration and specialisation process as a menace or as a source of advantages for the economic development and welfare of Europe is another matter.

The economic scenario at the regional level can be summarised by a threefold taxonomy: the old industrialised regions, characterised by a decline in their industrial base and the obsolescence of the skills of their labour force; the new industrialised regions, focused on innovative and fast-growing industries supported by an existing pool of human and financial resources; and the peripheral regions with no industrial tradition which — having experienced a brief period of growth and increased welfare in the golden age of the fordist production system (1960s and

[11] With the notable exception of the UK.

1970s) through the process of re-location and decentralisation of manufacturing — are now being pushed further away from the streams of technological and industrial innovation and development (Hilpert 1991).

After the crisis of the mass production model, the emergence of successful innovative industrial clusters, in the late 1970s, seemed to show that the opportunity still existed for technological development and lasting economic growth. Small and dynamic high-tech firms were regarded as the main, if not the sole, engine of economic development and regional innovation policy became the target of every public authority all over the world dreaming about the creation of a Silicon Valley clone. At that time the dream of a set of powerful regional innovation policies able to generate the development of any region seemed justified since the development of an innovative industrial cluster was promised simply by linking the existing pools of local resources to the dynamics of international supply and demand of innovative technologies, products and services (Blakely 1989). Nowadays, it seems widely established that it is rather difficult, if not almost impossible, for regional policies alone to radically change the local patterns of innovative production. The amenity of the place (for the peripheral regions) or the existence of an industrial tradition, for the old industrialised regions (Swann et al. 1998) could not constitute the basis of successful implementation of a modified "science-based" version of the Perroux's (1955) "growth pole" theory.

This chapter aims to reconsider the results coming from previous chapters of the thesis and to compare them with the main findings of the policy oriented literature in order to show the policy implications of the analysis and to develop a set of guidelines, if not an agenda, for policy makers interested in the location of high-tech firms and its relation to the process of regional development.

Before dealing with these matters, it may be useful to briefly list some *caveats* :

1. High-tech firm locations do not cause necessarily the growth and development of a region in terms of, income, employment, welfare etc.
2. There is no clear cut evidence on the superiority of a regional development strategy focused on the growth of existing high-tech firms versus an alternative strategy based on the entry of new high-tech firms.
3. Firms are not the only economic actors in the process of cluster development. Industrial clusters, or local systems of production, are collective actors (composed of firms and other institutions such as local authorities and professional associations) that can be targeted by specific economic policy interventions.
4. Firm location and cluster development are lengthy and cumulative processes where historical accidents, expectations, increasing returns and congestion play equally important roles.
5. National and regional peculiarities matter. What successfully worked in California, may not work as well as in Wales, Alsace, or Puglia.

This chapter is arranged into eight sections, which highlight the policy implications of the theoretical models and empirical evidence presented in the previous chapters of the thesis. The first section has an introductory function, the second explains the difference existing between geographic and agglomeration benefits in determining the location of firms and the development of a cluster, the third explicitly refers to geographical benefits as location factors, while the fourth section focuses on agglomeration benefits. The fifth section deals with the dichotomy between firms' entry and growth as determinants of cluster development, the sixth section is devoted to the analysis of the changing nature of a high-tech cluster during its stages of development and to the different policy instruments which are best suited to accompany and support each stage. The seventh section looks at science parks as an example of a policy instrument for supporting the birth and development of high-tech firms, while the final section summarises the results of the chapter through a series of guidelines for policy makers.

7.2 Geographical versus agglomeration benefits

All theoretical approaches to the location of economic activities have assumed (explicitly or not) that location benefits (i.e. the benefits which arise to an agent, firm or individual, from its particular location) are composed of two parts. The first part (geographic benefits) is dependent on the location and it is independent from other agents' behaviour; the second part (agglomeration benefits) is independent from the location and is solely dependent on the location choice of other agents. The relative importance attributed to geographic versus agglomeration benefits vary from one theory to another; consequently different theories involve different policy prescriptions (Rees and Stafford 1983).

Classical location theory and in particular the "least costs" approach (from Weber 1929 to Isard 1956) describes the firm's location decision as the minimisation of the sum of all transport costs associated with the distance existing between the input sources and the market[2]. An economic policy intervention aimed at developing a certain industry in a given location involves a two-stage intervention. The first stage is devoted to the identification (through direct interviews, sample surveys, econometric exercises) of a list of the "relevant location factors" for a given type of firms (i.e. high-tech ones). The second stage requires the implementation of specific policy interventions aimed at providing, in the chosen location, the previously identified factors (OTA 1984; Premus 1984).

Alternative theories[3] — from non price interaction models (Henderson 1977; Fujita 1986) to new economic geography (Krugman 1991a, 1995), from lock-in models (Arthur 1988) to industrial geography (Storper and Walker 1989), from biological models (Dendrinos and Mullally 1985) to the technological infrastructure approach (Feldman 1994) — focus on the influences that the number and type of already located firms have on the location decisions of potential new entrants. The consequences in terms of economic policy are connected to the design and implementation of a series of interventions aimed at regulating the dynamics of agglomeration economies and diseconomies.

In reality it is very difficult to distinguish between pure geographic and pure agglomeration factors since, apart from a very restricted list of physical characteristics which are connected to a particular location (such as climate, altitude, close-

[2] Agglomeration economies are treated as a third level effect arising from the intersection of isodipanes.

[3] Which directly or indirectly refers back to Marshall (1921).

ness to natural elements and amenities), several other elements — which are usually classified under the heading of geographic factors — may well be influenced by the location of other firms (i.e. transport infrastructures, whose efficiency and utility is crucially dependent on the number of users). On the other hand, the early or late emergence of congestive phenomena is heavily dependent on the local endowment of resources and inputs (which is an intrinsic geographic notion).

For these reasons in chapter 6 we have developed two distinct approaches which were able to highlight two different and complementary perspectives on the location process. The first perspective considers, in a panel-data framework, the number of located firms as function of a series of widely defined geographical factors. The second perspective considers geographical benefits as embodied in firms' location decisions and therefore looks at previous location as the sole determinants of new locations through time-series estimations.

The following two sections of this chapter will be devoted to the discussion of geographical versus agglomeration benefits as policy instruments to support firm's entry and clusters development. The view taken in this thesis is that interventions centred on locational benefits may obtain the greatest results in the early stages of development of a cluster, while the management of inter-firm interactions (agglomeration benefits) becomes the major task to be accomplished after the maturity phase of the cluster life cycle has been reached.

7.3 Geographical benefits and firms location

As stated above, the location decision of an individual firm, in this perspective, is conceived as solely dependent on "geographic" factors (i.e. independent of the previous location of other factors) and the task of a local authority, willing to encourage the innovative industrial development of an area, implies the provision of the location factors preferred by high-tech companies (Coleman and Jacek 1989).

These "preferred location factors" have therefore to be identified in some way. One of these methods, described in section 5.3.1, is via a direct survey questionnaire performed on a sample of firms, another (see section 5.3.3) involves the use of cross-sectional econometric analysis. The most famous example of a survey in order to detect the determinants of high technology plant location decisions is the analysis performed in the early 1980s for the US Office of Technology Assessment (Premus 1982). In this work a sample of 691[4] (out of the initial selected sample of 1750) firms, members of the American Electronic Association, were

[4] The sample was geographically very biased (even if the bias somehow reflected the spatial polarisation of high-tech firms in the period) since more than 300 firms were from California and around 200 were located in Massachussets. The sectoral structure of the sample was more representative of the spectrum of high-tech industries than one may suppose from the source of the sample (but with a remarkable bent for semiconductor and telecommunications which accounted for over 40% of the sample). A geographical breakdown of the responses allows the author to conclude that "if any aggregation bias exist, it is not serious" (Premus 1982, p. 26).

asked to rank a series of geographical[5] attributes on a four grades scale. The results showed that worker availability and skills, favourable tax structures, good educational institutions, low cost of living and efficient transportation infrastructure were, in order of importance, the most influential location factors.

An alternative but indirect way to identify the most relevant geographic factors which can explain the location decision of US high-tech firms has been performed in the original panel data exercise, presented in section 6.2 of the thesis. The advantages of such an analysis lie in the sample coverage, which is extended to the universe of high-tech firms operating in the US in 1986 and 1993. The drawbacks relate to the fact that such an analysis measures the degree of correlation between certain State's characteristics and the relative concentration of high-tech firms with severe limits concerning the direction of causality and the possible existence of spurious correlation.

According to the empirical results, States with a highly educated labour force (as shown by the high number of college enrolments), a strong technological infrastructure (as shown by high number of patents) and the presence of important metropolitan areas (as shown by the metropolitan population rate) are the most preferred locations for high-tech firms. Furthermore, State population and per-capita income are negatively (and significantly) correlated with the relative importance of high-tech firms, perhaps suggesting that innovative firms find an easier seedbed far from established productive centres (which usually are in rich and densely populated areas). State corporate income tax rate is also negatively related to the presence of high-tech firms.

When the same analysis is performed on the relative proportion of high-tech jobs, the list of significant regressors slightly change indicating that a dynamic (high number of business failures) and competitive (high export rate) economic environment with an innovative vocation (high number of patents) identifies a good location for high-tech jobs; while an old industrial region in crisis (as signalled by an high unemployment rate negatively correlated with the phenomenon at study) is a place to avoid. Other traditional location factors, often quoted in the empirical literature, such as the efficiency of transport infrastructures, the degree of unionisation of the labour force, and the level of local R&D seems to be not significantly correlated with the presence of a high level of high-tech employment.

These results seem therefore to imply that the policy interventions aimed at raising the presence of innovative firms in an area must raise the level of human capital in the local labour force (through an appropriate incentive structure for education), support the international competitiveness of the local economy, enforce the system of protection for intellectual property rights, and maintain low entry and exit costs. When implementing these policies, the local authority must be aware that the location process of high-tech firms is a very inertial process where "past success fosters future success" and where the economic effects of old innovative efforts affect more significantly the decision of entrepreneurs than cur-

[5] The exercise was conducted as a two stages choice. In the first stage firms were asked to rank the important factor to locate within a region (which in the US is composed by a number of states); in the second one firms were asked to rank the location factors relevant for the choice of a state location within a given region.

rent innovative efforts. The industrial structure of an area cannot be completely and sharply changed by policy interventions explicitly fostering the location of high-tech firms. On the contrary, the public authority, with its interventions, should focus on "already developed" areas encouraging the exploitation of locally available resources. For these reasons, the location of high-tech activities seems almost unsuitable as an instrument for narrowing geographical gaps in income, employment or other macro-economic variables.

7.4 Agglomeration benefits and firms location

The logistic models proposed in section 4.2.3, despite some empirical limitations, offer a useful criterion for classifying the large number of different policy interventions which, directly or indirectly, target the location of high-tech firms and the development of a local high-tech cluster. In brief all policies can be grouped into a threefold taxonomy which is based on the specific targeted parameter of the model.

For this purpose one can reformulate equation (4.5a) as follows:

$$\frac{dn_q}{dt} = r_q(n_q + \gamma_q)\left(1 - \frac{n_q + \gamma_q}{K_q}\right) \tag{7.1}$$

Where the growth of the high-tech cluster $\frac{dn_q}{dt}$ is modelled as a logistic function of the number of incumbent firms n_q; r_q is the incipient rate of growth (i.e. the maximum rate of growth which can be achieved by the cluster throughout its development); γ_q is a parameter which shifts the function along the n_q axis and represents the extent of critical mass phenomena; K_q is the maximum dimension of the cluster (i.e. the maximum number of profitable incumbent firms).

The only difference between equation (7.1) and equation (4.5a) lies in the introduction of the parameter γ_q which is intended to represent the relevance of a critical mass in the development process of an innovative industrial cluster and the possibility of implementing specific policy interventions aimed at overcoming this problem.

By referring to equation (7.1) we can therefore distinguish between three main types of policy interventions alternatively aimed at:

- overcoming the regional critical mass (by promoting the exogenous location of γ_q initial firms),
- increasing the maximum rate of regional growth (by increasing r_q),
- raising the long run equilibrium size of the cluster (by increasing K_q).

Let us now considering in details these three types of policy interventions[6].

A γ-type policy is designed to overcome the problem of the initial critical mass. Such a policy consists of a series of temporary financial and/or fiscal interventions which exogenously lower the location (or entry) costs in the region for a limited number of firms $(n_q = \gamma_q)$ in order to reach a level where the positive feedback dynamics of agglomeration economies can start to develop. A γ-type policy is therefore a kind of geographical benefits intervention to be used in order to foster the initial phase of development of a cluster in an "hostile" environment. It must be implemented when the targeted area is lacking any previous economic development, either in a particular industry (i.e. there have never been high-tech firms in this site) or in an absolute way (there have never been industrial firms in this site).

An *r*-type policy is designed to increase the positive externalities which are endogenously generated by the location of a new firm in the region. The intrinsic rate of growth, r_q, expresses the largest possible "attraction and generation" power[7] of a given number of located firms and influences the speed of growth of the cluster. An *r*-type policy explicitly supports the role played by agglomeration economies and knowledge spillovers in the development process of an high-tech cluster. The parameter r_q expresses also the difference between firms' birth and mortality rates in the region $(r_q = \beta_q - \delta_q)$. In this perspective, the concept of "attraction and generation" power of a cluster must be enlarged in order to take into account the early phases of firms' development within the region (the so-called incubation period, empirically estimated to last around 2 - 3 years (*Regional Studies* 1994), when the bankruptcy of a firm is most likely). An *r*-type policy can therefore aim at increasing the birth rate, and/or at decreasing the firms' "infant mortality" rate, within the region through appropriate interventions (such as innovation diffusion supporting policies, start-up incentives, provision of business planning services, diffusion of venture capital activities, etc.).

A *K*-type policy is designed to increase the regional "carrying capacity" which is the region's ability to sustain a given number of profitable "representative"[8] firms. Since the carrying capacity is a function of the local endowment of re-

[6] It is interesting to note that this threefold taxonomy of policies (γ, *r*, *K*) bears elements of similarity with the three "possible areas for policy interventions: (1) in the attainment of critical mass, (2) in the promotion of cooperative activities across firms and other institutions, and (3) in the coordination of investment decisions" identified by Temple (1998, p. 279).

[7] Which encompasses the entry of firms that were located outside the cluster and the "birth" of new firms inside the cluster.

[8] The concept of representative or average firms is introduced in the model to take into account the fact that, in reality, firms differ in size and that the growth of an high-tech cluster may imply either the increase in the number of established firms (i.e. the entry of new firms in the region) or the growth in size of a number of located firms. For a formal framework which explicitly models in different ways the entry and the growth of firms see Swann et al. (1998).

sources (inputs and infrastructures) and of the average level of use of these resources made by resident firms, then any public policy aimed at increasing the quantity and/or quality of local inputs and infrastructures, and at raising the efficiency of local firms can be defined as a *K*-type policy.

The desirability of these different development policies is crucially dependent on the preferred object of intervention, the chosen time framework for the implementation of the policy, the level of development of the targeted region, and the state and variability of the relevant external macro-economic environment.

As far as the target of the policy is concerned, *r*-type and γ-type policies are mainly addressed to firms, while *K*-type policies usually target the economic environment and the productive and urban infrastructures of the local economic system. According to this taxonomy r-type policies imply interventions such as start-up incentives, fiscal allowances, information diffusion programs. The establishment of a science park and the strengthening of the regional network of transport and communication infrastructures can be defined as *K*-type policies[9].

An alternative criterion, relates to the time horizon which is needed for the implementation of the economic policy interventions. Usually *r*-type and γ-type policies generate results in the short run, while *K*-type policy needs a longer time period to be effective. On the other hand, while the first two types merely influence the speed of development, *K*-type policies are the only ones capable of moving the cluster size from a lower equilibrium level to an higher one, thus ensuring higher sustainable long-run growth.

A third criterion refers to the stage of development of the targeted region. A γ-type policy can be implemented in an underdeveloped region without any industrial tradition. A *r*-type intervention is perfectly suited to be implemented in a "developing" region where the main problem is the establishment and survival of an initial core of high-tech firms. Finally a *K*-type intervention is designed to be implemented in an industrially developed region where competition on inputs and congestion of infrastructures are the main obstacles to the further development of the high-tech cluster[10].

A final criterion involves the state and variability (i.e. depth and frequency of shocks) of the relevant[11] external macro economic environment. According to macro-economic conditions the best development strategy (as developed in section 4.4) may involve pure *r*-type, or pure *K*-type policies when the environment is

[9] An example of γ-type policy is the re-location of governmental research institutions or public-owned firms.

[10] It is interesting to note that a famous US empirical study recognised that, in the early 1980s, *K*-type policies where already needed for supporting "high technology centres in the Silicon Valley and along Highway 128 (...) which, (despite their) formidable comparative advantage in the new high technology industries (...) will shortly approach the holding capacity of their respective regions" (Premus 1982, p. 18.)

[11] For an open and internationally integrated region the relevant external environment is the world, for a closed and underdeveloped region the relevant environment is limited to the nation, for an intermediate type of region, the external macro-environment is a group of countries.

stable; an intermediate policy when shocks are limited; and a mixed policy (i.e. a weighted combination of pure types) when shocks are deep and infrequent.

7.5 Entry versus growth policies

The empirical exercise of section 6.3 suggests that — although agglomeration economies are significantly and positively related with the high-tech specialisation of an area — scale economies play a dominant role in explaining the geographical distributions of high-tech activities (with the exception of sector 330, Instruments, where agglomeration economies are dominant) in four major industrial countries (US, UK, France, Italy) at the FLA[12] level. The results also confirm that local inter-industry spillovers at the FLA level are more important than intra-industry externalities at the SLA level. This is to say that the geographical specialisation of a FLA can be better explained in terms of a general high-tech vocation of that area than in terms of a specific industry specialisation of the larger area (SLA). There are many explanations for such a phenomenon: the first indirectly refers to locational shadowing (Arthur 1990) caused by the existence of a "best" location (FLA) which diminishes the relative attractiveness of alternative locations in the same SLA. The second explanation explicitly concerns the fact that firms belonging to the same group of industries (such as the high-tech sector as a whole) can generate a higher amount of location net benefits through a higher amount of inter-industry technological externalities and a lower degree of local competition on specific inputs. This result suggests that, since multi-technology clusters[13] are long-lived, they provide, in the long run, an higher degree of geographical specialisation of a particular industry. Taken together, these results seem to indicate that high degrees of industrial specialisation, measured by high value of the FLA employment location quotient, can be achieved by local policy intervention aimed at fostering a relative specialisation within the high-tech industry and at supporting the growth of size of incumbent firms.

However it should be considered that if geographical specialisation on its own makes the region more vulnerable to industry-specific shocks, geographical specialisation with a large average firm size raises the potential risk of bankruptcy of the entire local economic structure which depends entirely on few independent firms[14]. Furthermore it should be considered that a region with a given carrying capacity (or with a limited growth rate) can be saturated either by an increasing number of firms or by the growing size of the incumbents and that the two processes require very different economic and institutional conditions. This last observation highlights that the policy maker faces an intrinsic trade off between growth

[12] FLA, First Level Areas are defined in the thesis as the lowest considered level of analysis (US States, UK Counties, French *Départments*, Italian *Province*), while SLA, where S stands for second, are aggregation of FLAs (US Census divisions, UK, French and Italian regions).

[13] In accordance with the empirical evidence shown by Swann (1998).

[14] As witnessed by the historical case of the "rust belt" in the US.

policies and entry policies in order to foster the process of local industrial development (Markusen 1987).

The relevance of this industrial policy dilemma (entry versus growth) is further stressed by the empirical exercise of section 6.6[15] which has tested the empirical validity of three different explanations of the growth of high-tech clusters, which may be respectively attributed to Arrow-Romer, Porter and Jacobs. The analysis shows that the growth of industry employment in a given state is positively correlated with the local degree of sectoral specialisation (thus confirming the correlation between specialisation and growth) and negatively correlated with the degree of local monopoly (as measured by the average firm size). The empirical results seem therefore to support the story put forward by Porter in which both industrial specialisation and local competition foster the growth of an high-tech cluster.

Chapter 6 offers some alternative explanations of the fact that the actual degree of innovative industrial specialisation of an area is heavily dependent on the presence of large establishments (which enjoy higher economies of scale), while the growth rate of its sectoral employment is fostered by the presence of a competitive environment where small firms are predominant and economies of agglomeration play a major role. These interpretations — together with the ecological model of section 4.4 and the empirical evidence presented in section 6.3 — seem to suggest that policy makers must choose an industrial policy aimed at sustaining either firm entry or firm growth, according to the stage of development of the cluster and according to the prevailing macro-economic condition in the relevant external (national or international) economic environment.

However it seems generally acknowledged that, since small firms have lower entry and exit costs, they are best suited for the early stages (when the cluster is still competing for the lead in the industry) and the very late phases of development of a cluster (when the main core of the cluster/industry is suffering from its first crisis, and profit opportunity are reserved for niche productions only), while large firms perform better in an intermediate and mature phase (a rationalisation period which follows the emergence of the cluster as the national, or even international, leader in the sector and precedes the crisis in the cluster).

The issue of "optimal firm size" within an innovative cluster is, although indirectly, related to the role played by inter-industry technological complementarities (i.e. positive externalities arising from the location in the same site of firms in other high-tech sectors). It seems easier, for a policy maker, to introduce some degree of industrial diversification in a quasi single-technology cluster by fostering the entry of a number of small firms — belonging to other industries — than to obtain the same result by influencing the growth path of different firms in the cluster.

Finally, the positive and significant coefficients of the sectoral specialisation of a SLA in the regression equation explaining the sectoral specialisation of a FLA (in section 6.3) imply that the probability of establishing a successful high-tech cluster in a given location is heavily dependent on the industrial vocation of the larger surrounding area. This last result seems therefore to add further strength to

---

[15] Which — for reason related to the availability of a long time series of data — has been conducted on US data at the State level for the period 1975 - 1995.

the conclusion that the creation of an high-tech cluster cannot be easily used as a policy instrument for developing laggard and underdeveloped regions.

7.6 The nature of the cluster (and its policy implications)

If one agrees that the prime aim of economic policy in an advanced industrial country is to sustain its international competitiveness, and that the competitive advantages of countries are mainly based on their stock of knowledge — embodied in the skills and educational attainments of the labour force and incorporated in their technological infrastructures — then one must also agree that the leading objective of economic policy is to foster the development of the country's knowledge-based resources and to sustain its translation into competitive performance (Storey and Tether 1998a and 1998b).

In such a framework the traditional twofold structure, composed by science policy — which refers to the creation of new knowledge — and technology policy — which refers to the transformation of knowledge into products and processes and to the diffusion of innovations —, must be enlarged in order to accommodate a third type of policy, which refers to the trade in innovative goods and services, the adoption and adaptation of information and knowledge (both tacit and explicit), and internalisation within the existing local technological and economic system. This third type of policy — where science, technology and industry interact within a given institutional and territorial framework — calls for a series of economic policy interventions explicitly designed for supporting the generation and the development of innovative clusters (Temple 1998).

In the empirical literature explicitly devoted or even related to the analysis of such interventions (Preer 1992; Hilpert 1991; Luger and Goldstein 1991; Scott 1993; Castells and Hall 1994; Storper 1997) there is little if no reference to any theoretical reason justifying the need for a specific policy for innovative clusters.

A notable exception is represented by Temple (1998), who describes what he calls "industrial clusters" as a geographical concentration of specialised firms where quasi-co-ordination is achieved through mechanisms such as "information pooling and the consequential formation of shared perception regarding technological opportunities and their associated risks" (ibid., p. 272). On the same page it is stressed the role played by "shared perception" which is tentatively defined as an "impure public good (or 'club good') because use can only be made of it if at least some economic resources are committed to interpreting and refining it for the particular circumstance in hand" (ibid.).

According to Temple, on the basis of "at least one element of a 'public good', namely its non-rivalry" (ibid., p. 279), it is necessary to design and implement appropriate policy interventions to solve the inefficiencies caused by market failures when dealing with different aspects of the process of creation and development of innovative clusters. Temple identifies "three possible areas for policy interventions: in the attainment of the critical mass, in the promotion of co-operative activities across firms and other institutions, and in the co-ordination of investment decisions" (ibid.).

This stimulating work needs further extensions and refinements. Several characteristics — such as the existence of a critical mass, mutual trust, joint use of a common pool of highly mobile human and technological resources, technological and productive interdependence which is implied by inter-firm specialisation and division of labour, and the definition and setting of technical standards — of an innovative industrial cluster, besides the "shared vision", call for a definition of the cluster *per se* as a peculiar type of public good. The first peculiarity lies in the coincidence of producers and users. The public good "innovative industrial cluster" is a by-product of the autonomous firm's decision to produce a private good in a given site[16]. The second peculiarity refers to the fact that the "innovative industrial cluster" seems to change its nature according to the number of users[17].

If such are the features of an innovative industrial cluster, then it necessarily follows that the development of innovative industrial clusters is a case not only for public provision but also for interventions aimed at regulating its use. In order to discuss this issue it may be helpful to make use (with some changes) of a theoretical framework proposed, among others, by Adams and Mc Cormick (1987) and presented in table 7.1, together with figure 7.1[18].

Figure 7.1 Agglomeration costs and benefits and critical sizes of a cluster

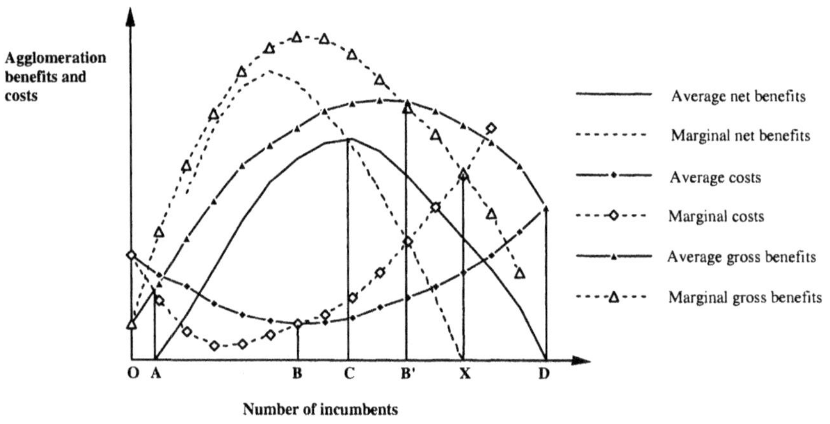

[16] This is a case of joint production which can be described as a pure production externality.

[17] For the existence of a cluster, a minimum number (the critical mass) of users (producers) is needed. After that number is reached, the good is not only non-rival but the more users (producers), the higher the level of per-user benefits. However, once a second and higher level is reached, the good becomes congestible (i.e. rivalry begins to appear) and, if exclusion is not feasible, the good — or better the benefits generated by its use — disappears.

[18] Already shown in chapter 4, as figure 4.2.

Table 7.1 A taxonomy of goods

Consumption		Exclusion: Feasible	Exclusion: Non feasible
	Rival	**1** private goods	**4** commons **Innovative cluster phase III**
	Congestible	**2** club goods **Science park phase II**	**5** non marketable impure public goods **Innovative cluster phase II** ↑
	Non- rival	**3** marketable public goods **Science park phase I** ↑	**6** non marketable public goods **Innovative cluster phase I** ↑
	Unprofitable or inexistent	↖ Potential location ↗ as science park and as innovative cluster **0**	

Lets start from figure 7.1, which shows marginal and average[19] agglomeration benefits (costs) which are enjoyed (incurred) by a firm which locates in an area (which will be later defined as either "science park" or "innovative cluster" according to the feasibility of exclusion). It is now possible to follow the evolution of the area as more and more firms decide to locate there (and the dimension of the area grows from **O** to **D**); the area will change its nature by going through the different cells of table 7.1 following alternatively the right or the left path (column).

The two columns of table 7.1 refer to the feasibility of exclusion. It seems reasonable to assume that, in general, it is not possible to exclude a specific entrant firm willing to join an innovative industrial cluster without violating the most basic rules of economic democracy[20]. The situation is completely reversed when the cluster under study is a property-based initiative (such as a science park) and, as such, it is explicitly defined as a "club" in which any new membership must be approved by the incumbents (or by an *ad hoc* committee).

The two typologies of innovative area have in common an early development stage $(\mathbf{O} < n \leq \mathbf{A})$, characterised by strong indivisibility, when the number of located firms is smaller than the critical mass. In such a situation, the area — despite the fact that both marginal and average net benefits are increasing — is a mere potential site for the establishment of an innovative cluster or a science park since the average agglomeration benefits are zero or negative (i.e. these benefits are equal or lower than those enjoyable outside the area[21]). After this initial stage the two development paths diverge.

[19] I.e. the benefits available to each incumbent firm.

[20] And once the firm is located in the area is even more difficult to exclude it from enjoying most of the benefits.

[21] This benchmark level of benefits is assumed exogenously fixed and for convenience is set equal to zero.

The development path of an innovative cluster, in which exclusion is unfeasible, is divided in three distinct phases. In the first phase $(\mathbf{A} < n \leq \mathbf{C})$, which follows immediately after reaching the critical mass, the cluster behaves as a pure non marketable public good. Each firm produces its output (private good) and indirectly produces a positive externality which is directly enjoyable by every firm located into the cluster. At this stage the level of externality is an increasing (concave) function of the number of local firms. In the second phase $(\mathbf{C} < n \leq \mathbf{X})$ the innovative cluster acts as a non marketable impure public good. The incumbents (i.e. the firms already located in the cluster) would like to restrict any entry, since at **C** the average agglomeration net benefits are maximised and any further entry reduces them. A simple welfare analysis show that entry should continue (from a cluster social planner's point of view) until the regional industrial mass reaches **X**, where marginal costs equal marginal benefits. However the entry process continues, driven by the existence of positive average benefits and, once **X** is exceeded, the cluster initiates its third phase $(\mathbf{X} < n \leq \mathbf{D})$ becoming a common resource which is inevitably bound towards over-exploitation. Finally the net entry process ends in **D** when the excessive entry of firms drives the agglomeration benefits to zero[22]. In table 7.1 the evolution of the cluster is graphically represented by a path which, starting in cell 0, gradually reaches cells 6 and 5 and finally settles in cell 4.

A different development process is followed by a managed (property based) form of industrial cluster such as a science park. In the first phase (from **A** to **C**), the science park behaves as a marketable public good which, thanks to the feasibility of exclusion, is provided by private operators. After that, the science park can be managed as a club in order to control the effects of congestion in a way which allows it to reach the optimal social dimension **X**. Once this dimension has been reached, the park's management acts to keep it stable and further entries are allowed only to compensate in the event of exits or bankruptcies. In table 7.1 the evolution of a successful science park is graphically represented by a path which, starting in cell 0, gradually reaches cells 3 and finally settles in cell 2.

From the above description, it necessarily follows that policy instruments for innovative areas must be chosen according to the nature of the area and to its development stage. It is in fact very different — apart from an initial phase, where the common problem is to reach the critical mass — managing agglomeration dynamics with and without the possibility of exclusion.

In the case of an innovative cluster, where it is impossible to limit entry, local policy interventions can only indirectly regulate the entry speed[23] and, above all, provide factors which exogenously determine the carrying capacity of the cluster. In particular local authorities can reduce the congestion pressure on a given cluster by improving the amount of locally available resources and infrastructure (i.e. enlarging the existing road system, changing the destination of state-owned land, increasing the local level of human capital, raising the local level of supply of spe-

[22] Even though new entries are still possible, at the expense of incumbents.

[23] A reduction in the entry speed can facilitate the adjustment of supply to changes in the local demand of goods and services.

cialised services, etc.) or by virtually enlarging the geographical dimension of the cluster by lowering the transportation costs from outside the area. A successful example of such a policy (and of co-operation between local public and private institutions) is represented by the California MICRO program which throughout the Eighties provided funding for graduate fellowships and faculty research projects, by matching grants from private industry (OTA 1984). Another example of local policy[24] has been implemented during the Nineties in Silicon Valley and has indirectly targeted the extremely high cost of housing, which started to emerge as a serious barrier to the further development of the cluster. In this the transportation network has been improved through the enlargement of existing highways and the construction of new ones and through strong local support for the government-led deregulation of the airlines industry, which triggered the development of cheap airline "commuter" services to nearby states (Padmore and Gibson 1998).

National policies can positively influence the creation of new innovative clusters by reducing local congestion in already established ones. In this way a sort of equilibrium à la Tiebout (1956) can be achieved by offering a spatially differentiated and wider supply of locations to the population of high-tech firms. The most famous case of national policies is represented by the Japanese Technopolis Programme which, officially started in 1983 by national law, planned the establishment of a network of 26 high-technology production centres in order to develop local innovative R&D capacity and to trigger the development of such industries locally (Smilor et al. 1988; Gibson et al. 1992). This program intends to spur high-tech industries over the country through an initial pump-priming subsidy to stimulate the development of local R&D capacity in two ways. Firstly through the relocation, from the congested metropolitan areas, of existing high-tech firms, and second through assisted self-development of existing local industries (i.e. diffusion of innovative technologies into traditional sectors) (Castells and Hall 1994).

A third, hybrid, but very common, type of intervention, composed of local policies performed by national authorities, is exemplified by procurement policies. This policy instrument, in which government plays a major role in local markets as purchasers of goods and services, has been a major element of success in the development of high-tech industries in Sweden and in the US. In Sweden, carefully structured State purchases led to the development of industrial strength in power transmission and rail transport (Edquist 1995). In Silicon Valley and in the Route 128 area, the research contracts and purchases by the Department of Defence and NASA acted as powerful subsidies of local R&D allowing the development of innovative technologies which were the riskiest in terms of investment. Furthermore, the practice of "second sourcing" and technology sharing — together with the public diffusion requirement of the discoveries[25] realised through the Department of Defence funding — led to a rapid technology diffusion among firms and a high rate of formation of spin-off firms (Preer 1992; Scott 1993; Saxenian 1994).

[24] Although mainly local in its effects this can be better defined as an example of interactions between local and national policies.

[25] With the exception of those classified as "military secrets".

In the case of science parks, the presence of an explicitly designed management function should allow, in principle, a decisive influence and control on the evolution of the area through a policy of selective admission and entry fees. Since this issue has generated an impressive amount of literature[26], it seems sensible to devote the following section to the analysis of the phenomenon in order to summarise the empirical findings and to organise them in a tentative taxonomy.

7.7 Policy instruments in actions: science parks and technopolis

When surveying the literature on policy instruments devoted to supporting the establishment and development of innovative industrial clusters it is impossible to avoid a discussion about strengths and weaknesses, pros and cons, cost and benefits of science parks and technopolis.

Let us start with a definition. A science park, or technopolis, is officially defined as: "A property-based initiative which: has operational links with Universities, Research Centres and/or other Institutions of Higher Education; is designed to encourage the formation and growth of knowledge-based industries or high value-added tertiary firms, normally resident on site; has a steady management team actively engaged in fostering the transfer of technology and business skills to tenant organisations" (Rowe 1988)[27].

Science parks are generated by formal organisations which sell or lease land and/or building to firms and other organisations whose principal activity is basic or applied R&D and/or the development of new products or processes. This definition therefore excludes other forms of high-tech clusters, centres or corridors (such as Route 128 and Silicon Valley in the US, the M4 corridor in the UK, the *Région Parisienne* in France, the metropolitan area of Milan in Italy) where high-tech firms have clustered without and outside formal property-based organisations. It is however interesting to note that some of these "spontaneous" clusters have originated from a science park (the most famous example being Silicon Valley which emanated from the Stanford Research Park in Palo Alto).

A tentative taxonomy of science parks is based on their relationships with the local economic environment. According to this criterion, the main types of science parks refer to three different regional and urban management functions: development, re-conversion and de-localisation (Maggioni 1990 and 1995). The first type refers to science parks (such as Research Triangle Park and Utah Research Park in the US, Cambridge Science Park in the UK, CSATA in Italy) which are located in areas of little, or no, established tradition in manufacturing, willing to build their economic development on the new technology-based industries. The second refers to science parks (such as Nancy Brabois and Metz 2000 in France, Aston in the UK, Tecnocity in Italy) located in areas characterised by the strong presence of

[26] Mainly empirical and based on case-histories.

[27] This definition is can also be found in the official WWW site of IASP, the International Association of Science Parks (http://www.iaspworld.org).

one or a few traditional sectors[28] which are now in crisis; the third refers to science parks (such as Sophia-Antipolis in France, Tsukuba in Japan) whose main objective was to relocate part of the scientific and technological resources which were gravitating around a metropolis in order to reduce the existing urban and infrastructural congestion.

Each type of science park uses different tools and incentives in order to promote the development of the area according to the prevailing conditions of the local economic system. Development science parks aim to build an entrepreneurial attitude and culture within the academia and provide strong incentives to the creation of spin-off firms from university laboratories. Re-conversion science parks underline the value of continuous education and training of the labour force and promote the diffusion of new technologies into mature sectors. De-localisation science parks attract scientists and highly qualified workers by offering better living conditions and try to substitute the geographical marginality of the area by the availability of high level technological infrastructures and a proximity to international airports.

All these typologies share a common feature: they are based on the coexistence (and possibly, co-operation) of three different actors: firms[29], public authorities, and research institutions. To be effective in a science park, these three actors, despite their different aims, must agree on the targets to be set, the tasks to be accomplished and the instrument to be used. To be efficient in a science park, each of the three actors must specialise in the provision of specific resources according to its own comparative advantages (Gibb 1985).

In particular, firms — which should develop strong producer-user interactions — are the main resources of a science parks, since they provide the co-ordination between capital and labour in new and advanced sectors. The main objective for a firm based in a science park is to enjoy the agglomeration economies — in terms of knowledge spillovers, reduced search costs, availability of skilled labour costs and specialised intermediate inputs — and to establish fruitful relationships with basic and applied research institutions. Luger and Goldstein (1991) suggest a twofold taxonomy of science park development strategies (exogenous versus indigenous) related to the prevalent type of firms. Exogenous development means that the science park is "initiated, propelled and controlled by organisations located outside the region" (...) (while) indigenous development refers to development that is regionally initiated and planned" (ibid., p. 18). In the first case, the science park promotes the location of R&D departments and branch plants of large national and foreign multinational corporations; in the second a central role is played by locally owned small and medium enterprises.

Public authorities (both local and national) see the science park as a flexible and systemic tool to implement innovation supporting policies. They may act according to two different "styles" of policy. In the "early American style" public authorities play only an indirect role on the development of science parks, acting mainly as promoters of research contracts and purchasers of high-tech goods and

---

[28] Such as: Ferrous metals, Fabricated metal products, Motor Vehicles.

[29] Which are mainly private-owned, but — in the French and Italian experience — may also be public-owned ones.

services. In the "European[30] style" — in which the science park is also used as an instrument to narrow existing geographical gaps by developing peripheral areas — the government plays a direct role in the establishment, and sometimes also in the management, of the park. The main objectives of public authorities, involved in a science park, span therefore from increasing the international competitiveness of the national economy in knowledge based sectors, to the reduction of existing intra-national economic and social unbalances, from the diversification and revitalisation of the local economic structure, to the reduction of public expenditure for universities.

Universities and research institutions provide the scientific engine of science parks. They are responsible for the education of the local labour force, the monitoring of the international scientific and technological dynamics, the expansion of scientific and technical knowledge, and the diffusion of innovations within the local industrial structure. Their main objective when joining a science park is to achieve and maintain international scientific reputation, to develop basic research with potential industrial application, and to encourage funding support from private firms.

A science park can therefore be seen as a managed form of innovative industrial cluster where the achievement of the critical mass, the promotion of co-operative activities and the co-ordination of investment decisions are granted and regulated by specific policy actions. The achievement of the critical mass, in the early phases of development of a science park, is accomplished through the location (or re-location) of public research laboratories and/or public owned firms, the promotion of academic spin-off and the attraction of subsidiaries of multinational firms. The encouragement of co-operative activities is the main task of the science park's management which is actively engaged in fostering the interaction between firms and other institutions located in the park and is also fostered by specific national and super-national research projects[31]. The co-ordination of investment decisions is often attained within the science park through the setting of technical standards and the existence on site of an active financial sector especially devoted to the funding of innovative activities (e.g. venture capitals etc.).

As for any policy measure, for science parks it is also important to establish not only some guidelines for their design and implementation but also to list a series of performance indicators which should be used in order to evaluate success or failure. Such indicators have been explicitly and specifically designed for science parks; however, some of them can be easily adapted and used in order to evaluate the performance of any innovative industrial cluster. General indicators are: the number of new jobs created since the establishment of the science parks, the number of new firms generated and the number of already existing firms which relocated in the park, the number of patent registered by firms and other institutions resident in the park. Specific indicators are: the number of firms which have used services provided by the science park's management, the amount and source (public, venture capital, banking credits) of funds which have been raised and invested in the park.

---

[30] Or "late American style".

[31] This is the explicit scope of several UE programs such as Eureka.

At present, science parks are established policy instruments only in the US (where they originated in the 1950s) and in France (late 1960s). In other European countries, science parks and technopolis were only developed in the last ten — fifteen years and therefore it is more difficult to value their efficacy. A number of empirical studies for the UK and the Netherlands (Westhead and Storey 1994; Dollar and Wolff 1993; Sternberg 1997) show that science parks have a very limited positive influence on the rate of growth (measured both on employment and establishment) and on the survival rate of tenant firms. The situation is thus rather different from the American and the French experience where studies shows either a marked general positive effects for the US case (Luger and Goldstein 1991) or, at least, an industry-specific positive effect in France (Longhi and Queré 1991).

7.8 Guidelines for policy makers

The theoretical models presented in chapter 4 and the empirical evidence presented in chapter 6 of the thesis have underlined the path dependence and non ergodic features of the development process of high-tech clusters. The previous sections have discussed in depth some of the issues relating to the choice of the best policy intervention for promoting the generation and sustaining the development of high-tech clusters.

It is rather difficult to summarise all these findings down to a sketchy list of policy prescriptions. For this reason, before trying that, it seems sensible to highlight some of the major issues involved.

The first issue refers to the difficulty in identifying dynamic new sectors of production before they have started to grow and "yet once growth has begun and agglomerative forces have set in, the chances of creating a new agglomeration elsewhere are much diminished, for the first-mover advantages and localised externalities that accrue to the original case help to crowd out late imitators" (Scott 1993, p. 256).

Another issue relates to the degree of spatial competition. Sometimes the strategies implemented by competing local public agencies to develop an high-tech cluster in their own jurisdiction (by offering similar locational benefits) may generate an "arms race" type of game in which costs are far in excess of the benefits. The crucial assumption for the Tiebout (1956) principle to hold is, together with perfect mobility of firms, the existence of a sufficient degree of differentiation between localities, so that firms can locate according to their valuation of different mixes of locational factors. Furthermore, it must be considered that regional competition mechanisms are likely to widen rather than narrow existing regional disparities (Kaldor 1970), since less developed regions are often unable to offer conditions and incentives for attracting high-tech firms comparable to those provided by more advanced regions.

A further issue deals with the optimal level of implementation and funding of policy interventions supporting the establishment and development of an innovative industrial cluster (whether in a spontaneous or managed form). In particula the main option is between the local or the national level[32]. The answer, which from an institutional perspective is influenced by the degree of centralisation of the different countries, from an economic point of view, is related to the spatial distribution of assigned tasks and expected benefits.

A final issue concerns the statement that regional policies in general (and innovative cluster interventions in particular) cannot be focused — as they were conceived in the '60s and early '70s — on the geographical redistribution and equalisation of income and employment. The difficulties of implementing radical local economic transformation and of establishing new agglomeration *ex-nihilo* are widely known and it is also well recognised that the only viable alternative is to build incrementally on what already exists. This is not to say that underdeveloped regions should be left over; on the contrary anywhere there are (even limited) local endowments of human and capital resources, there are real possibilities of building a development process. It is however clear that industrial innovative clusters cannot be started everywhere and that, in the case of an absolute lack of initial advantages, only massive, concerted and extremely expensive actions by central government authorities have the possibility of producing results whose net effects are, however, ambiguous.

At this stage one may summarise the main lessons which can be extracted by distilling the complex of theoretical models, empirical evidence, and case histories contained in the thesis in a series of sketchy policy suggestions[33].

A policy maker willing to design and implement a series of policy interventions aimed at fostering the creation and promoting the development of an innovative industrial cluster must always:

1. Identify the main type of locational instrument which will be used. In particular, choose between innovative cluster and science park models because they are different and require different policies.
2. Choose the best development strategy according to the different development stage of the area, and the level and variability of the relevant macro-economic environment outside the area.
3. Take into account the path dependent and inertial nature of the location process, trying therefore to marginally modify rather then completely change the industrial location pattern of an area[34].
4. Consider carefully the demand and supply of innovative inputs, not only in the targeted area but also in the neighbouring ones (with special reference to those

[32] For the European countries the level become threefold because, especially after the White Paper (Delors 1993), one must consider the European Union as a major player in any innovation related policy.

[33] Which is partially modelled on Castells and Hall, 1994.

[34] In situation where there is a "considerable concentration of 'institution and networks' in a central core region it may be best to attempt at local decentralisation to the periphery of this region rather than developing distant region. (...) If (however) it is desired to build up synergistic capacities in other, more peripheral region, it will almost certainly be necessary to concentrate in one or two target areas that offer the best prospects (...). A scattershot strategy, though more acceptable in political terms may fail to build up momentum anywhere (Castells and Hall 1994, p. 248).

regions which have long established relationship with the targeted area) in order to avoid wasteful duplication of efforts and potentially inefficient competition[35].

5. Choose between an exogenous and an indigenous development strategy according to the actual potentiality of the area (Luger and Goldstein 1991), remembering that "branch plants are better than no plants" (Castells and Hall 1994, p. 248).
6. Choose between a multi-technology versus a single-technology type of cluster by taking into account the expected degree of convergence of intra and inter industry technologies and the different development time-paths of the two types of cluster, since the former grows more rapidly while the latter is more long-lived being less prone to congestion (Swann 1998).
7. Consider that the time horizon in which an innovative industrial cluster reaches maturity from its birth is on average longer than the usual period in which a firm expects to reach the break-even point of an investment and also longer than the usual political election cycle. For this reason it is essential to provide a series of protective mechanisms against premature accusation of failure. Conversely one should also avoid the temptation to expand the scope and the scale of the experiment in order to escape from evaluation processes.

[35] A super-regional coordination mechanism would have avoided the bidding war which developed in the early 1980s between a number of states and communities (including Research Triangle in North-Carolina, San Diego in California, Phoenix in Texas) for the location of the research headquarters of the semiconductor industry's Microelectronic and Computer Technology Corporation (MCC).

Chapter 8 - Conclusion and research agenda

> This need for the emergence of "growing points" or "growth poles" in the course of the development process means that international and inter-regional inequality of growth is an inevitable concomitant and condition of growth itself. Thus in the geographic sense, growth is necessarily unbalanced.
>
> A.O. Hirschmann (1958) *The Strategy of Economic Development.*

The thesis has looked at the location process of high-tech firms and at the emergence of spatial industrial clusters. The main contributions achieved by the work are the following.

1. Clustering (i.e. the spatial concentration of high-tech industries in a given location) is a relevant phenomenon in four major industrialised countries. Clustering is not limited to innovative industries but, in general, these sectors show an above average bias towards spatial concentration. This has been checked with reference to two "benchmark" sectors (Motor vehicles and Textiles) which have been signalled by Krugman (1991a) as two clustered traditional industries.
2. The extent of industrial clustering is heavily dependent on the spatial and industrial definitions of the measurement unit. For this reason the statistical analysis has been conducted at two different geographical levels and a careful harmonisation of different national industrial classifications has been performed. Clustering is also heavily influenced by the variable chosen for the analysis. In the thesis both employment and establishments data have been used. Finally, different concentration and inequality indexes may give different results.
3. The US appear to have the highest concentration ratio when clustering is measured in term of establishments. If (bearing in mind the usual caveats) one thinks of the US as a similar country to the future monetary and economic European Union, then high-tech clustering seems bound to increase in the next future in Europe.
4. Different streams of literature have been surveyed and used in the thesis to look at the relation between clustering and location processes. A useful twofold taxonomy of these contributions is based on the main source of locational benefits, these being either geographical (i.e. site-dependent) or agglomeration (firm-dependent) benefits.
5. Location models should explain why firms entry processes into the cluster takes time. The most interesting rationale refers to so called "rank" effects (i.e. the

existence of either subjective or objective differences in the set of potential entrants) and to epidemic effects (which may either refer to the diffusion of information or to an endogenous decrease in uncertainty).

6. Cluster development models should take into account the internal dynamics of a clusters development as well as the external ones. By "external" we mean three different sort of processes: the inter-industry relations within the same region, the inter-regional relations within the same industry, the influence of the surrounding national and international macro-economic conditions. When expectations are introduced into ecologically derived models of cluster development, these become very similar to the (subjective) rank effects modelled described in (v).
7. A highly educated labour force, strong innovative vocation, and low corporate income tax seem to favour the location of high-tech firms in dynamic areas. Established industrial centres and declining areas are, on the contrary, deserted by high-tech industry.
8. Scale economies, in general, outweigh agglomeration economies as the major determinant of industrial specialisation. Inter-industry technological externalities seem stronger than geographical knowledge spillovers. Furthermore, for the Instruments industry, there is empirical evidence of locational shadowing.
9. In general, competitive interactions between high-tech clusters are not symmetrical. The development of one cluster may well be negatively influenced by the industrial specialisation of a neighbouring area even if the development of the latter is uninfluenced by the industrial activity of the former.
10. Industrial specialisation and local competition are, according to Porter (1990) and to our empirical evidence (based on the evolution of 112 US high-tech clusters from 1975 to 1995), the "engines" of cluster growth.
11. The desirability of different types of policy is heavily dependent on the development stages of the cluster (within its life cycle) and the macro-economic conditions of the relevant external economic environment.
12. High-tech industrial clusters may be developed by policy makers either as innovative clusters or as science parks. The choice is crucial, since on it depends the future evolution of the areas and the different policy instruments which must be implemented in order to ensure their success.

Future research plans involve the extension of the database to other advanced countries (such as Germany, Canada and Japan) and the use of "spatial indexes" as explanatory variables of the international competitiveness of different sectoral and national systems of innovation; the re-estimation of the industrial and spatial interaction effects of section 6.5 at a lower geographical level, (i.e. different counties within high-tech clusters such as Silicon Valley, Route 128, Research Triangle); the collection of firm level data in order to test the diffusion derived model of section 4.5; and the construction of a simulation framework for exploring the spatial diffusion patterns of high-tech clusters when agglomeration economies and diseconomies, and spatial and industrial spillovers are present.

References

Acemoglu D. (1992): Labour Market Imperfections, Innovation Incentives and the Dynamics of Innovation Activity. Mimeo, London School of Economics, London

Acs Z.J., Audretsch D.B. (1989): Small Firm Entry in US Manufacturing. Economica, 56, 255-265

Acs Z.J., Audretsch D.B. (1992): Technological Regimes, Learning, and Industry Turbulence. In: Scherer F.M., Perlman M. (Eds.): Entrepreneurship, Technological Innovation, and Economic Growth: Studies in the Schumpeterian Tradition. University of Michigan Press, Ann Arbour, 305-320

Acs Z.J., Audretsch D.B., Feldman M.P. (1992): Real Effects of Academic Research: Comment. American Economic Review, 82 (1), 363-367

Acs Z.J., Audretsch D.B., Feldman M.P. (1994): R&D Spillovers and Recipient Firm Size. Review of Economic and Statistics, 100 (2), 336-340

Adams R.D., Mc Cormick K. (1987): Private Goods, Club Goods, and Public Goods as a Continuum. Review of Social Economy, vol. XLV, n. 2, 192-199

Ady R. (1986): Criteria Used for Facility Location Selection. In: Walzer N., Chicoine D. (Eds.): Financing Economic Development in the 1980s. Praeger, New York, 72-84

Allen P.M. (1976): Evolution, Population Dynamics and Stability. Proceedings of National Academy of Science of the USA, 73 (3), 665-668

Allen P.M. (1980): Self Organisation in Human Systems. Revue Belge de Statistique, d'Informatique et de Recherche Opérationnelle, 20 (4), 21-76

Allen P.M., Sanglier M. (1978): Dynamic Models of Urban Growth. Journal of Social and Biological Structures, 1, 256-272

Allen P.M., Sanglier M. (1981a): Urban Evolution, Self-organization and Decision Making. Environment and Planning A, 13, 167-183

Allen P.M., Sanglier M. (1981b): A Dynamic Model of Central Place System. Geographical Analysis, 13 (2), 149-164

Andersen E.S. (1994): Evolutionary Economics. Pinter, London

Antonelli C. (1988): I determinanti della distribuzione territoriale dell'attività innovativa in Italia. In: Antonelli C., Cappellin R., Jannaccone Pazzi R.: Le politiche di sviluppo locale. Angeli, Milano, 94-120

Arrow K. (1962): Economic Welfare and the Allocation of Resources for Invention. In: Nelson R.R. (Ed.): The Rate and Direction of Inventive Activity. Princeton University Press, Princeton, 609-625

Arthur W.B. (1983): Competing Technologies and Lock-in by Historical Small Events: the Dynamics of Allocation under Increasing returns. International Institute for Applied Systems Analysis Papers, 83-92, Laxenburg, Austria

Arthur W.B. (1988): Urban System and Historical Path Dependency. In: Ausubel J.H., Herman R. (Eds.): Cities and Their Vital Systems. National Academy Press, Washington D.C., 85-97

Arthur W.B. (1989): Competing Technologies, Increasing returns, and Lock-in by Historical Events. Economic Journal, 99, 116-131

Arthur W.B. (1990): Silicon Valley Locational Clusters: when Do Increasing Returns Imply Monopoly?. Mathematical Social Sciences, 19, 235-251

Aydalot P., Keeble D. (Eds.) (1988): High Technology Industry and Innovative Environments: the European Experience. GREMI-Routledge, London

Audretsch D., Vivarelli M. (1996): New-Firm Formation in Italy: a First Report. Economic Letters, 48 (1), 75-81

Bagnasco A. (1977): Tre Italie, la problematica territoriale dello sviluppo italiano. Il Mulino, Bologna

Baldwin R.E. (1997a): Global Divergence, Trade and Industrialisation: The Geography of Growth Take-Offs. Mimeo, Graduate Institute of International Studies, Geneva

Baldwin R.E. (1997b): Agglomeration and Endogenous Capital. European Economic Review, 43 (2), 253-80

Baldwin R.E., Martin P., Ottaviano G.I.P. (1997): Global Income Divergence, Trade and Industrialisation: the Geography of Growth Take-Offs. Mimeo, Graduate Institute of International Studies, Geneva

Banerjee A.V. (1992): A Simple Model of Herd Behavior. The Quarterly Journal of Economics, 107 (3), 797-817

Baptista R., Swann G.M.P. (1996): A Comparison of Clustering Dynamics in the US and UK Computer Industries. Paper presented at the 23rd EARIE Conference, Wien, September 7-10

Barkeley D.L., McNamara K.T. (1994): Do Surveys Produce Helpful Insights?. International Regional Science Review, 17 (1), p. 23-47

Bass F. (1969): A New Product Growth Model for Consumer Durables. Management Science, 15, 215-227

Becattini G. (Ed.) (1987): Mercato e forze locali: il distretto industriale. Il Mulino, Bologna

Becattini G. (1989): Modelli locali di sviluppo. Il Mulino, Bologna

Becattini G. (1998): Distretti industriali e made in Italy: le basi socioculturali del nostro sviluppo economico. Bollati Boringhieri, Torino

Beckmann M.J. (1968): Location Theory. Random House, New York

Beckmann M.J. (1990): Distance Inputs and Spatial Markets. In: Chatterji M., Kuenne R.E. (Eds.): New Frontiers of Regional Science. McMillan, London, 19-25

Beckmann M.J., Thisse J.-F. (1986): The Location of Productive Activities. In: Nijkamp P. (Ed.): Handbook of Regional and Urban Economics, 1. Elsevier, Amsterdam 21-95

Begg I.G., Cameron G.C. (1988): High Technology Location and the Urban Areas of Great Britain. Urban Studies, 25, 361-379

Beltrami E. (1987): Mathematics for Dynamic Modelling. Academic Press, Boston

Benko G., Lipietz A. (Eds.) (1992): Les régions qui gagnent. PUF, Paris

Bester H., De Palma A., Leininger W., Von Thadden E.L., Thomas J. (1991): The Missing Equilibria in Hotelling's Location Game. CENTER Working Paper, 15, Tilburg University, Tilburg

Bikhchandani S., Hirshleifer D., Welch I. (1992): A Theory of Fads, Fashion, Custom and Cultural Change as Informational Cascades. Journal of Political Economy, 5, 992-1026

Blakely E.J. (1989): Planning Local Economic Development. Theory and Practice. Sage, London

Bluestone B., Jordan P., Sullivan M. (1981): Aircraft Industry Dynamics. Auburn House, Boston

Boudeville J.R. (1966): Problems of Regional Economic Planning. Edinburgh University Press, Edinburgh

Boyer R., Saillard Y. (Eds.) (1995): Théorie de la régulation: l'état des savoirs. La Découverte, Paris

Bramanti A., Maggioni M.A. (Eds.) (1997): La dinamica dei sistemi produttivi territoriali: teorie, tecniche, politiche. Angeli, Milano

Brezis E.S., Krugman P. (1993): Technology and the Life Cycle of Cities. NBER Working Paper Series, 4561, NBER, Cambridge (Mass.)

Brittain J.W., Wholey D.R. (1988): Competition and Coexistence in Organizational Communities. Population Dynamics in Electronic Components Manufacturing. In: Carrol G.C. (Ed.): Ecological Models of Organizations. Ballinger, Cambridge (Mass.), 195-222

Browning J. (1980): How to Select a Business Site. McGraw-Hill, New York

Brülhart M., Torstensson J. (1996): Regional Integration, Scale Economies and Industry Location in the European Union. CEPR Discussion Paper Series, 1435, CEPR, London

Buswell R.J. (1983): Research and Development and Regional Development. In: Gillespie A. (Ed.): Technological Change and Regional Development. Pion, London, 9-22

Camagni R. (Ed.) (1991): Innovation Networks. Spatial Perspectives. Belhaven, London

Capelo A.C. (1989): Modelli matematici in biologia. Decibel, Padova

Carrol G.C. (Ed.) (1988): Ecological Models of Organizations. Ballinger, Cambridge (Mass.)

Castells M. (1985): High Technology, Space, and Society. Sage, Beverly Hills

Castells M. (1988): The New Industrial Space. In: Sternlieb G., Hughes J.W. (Eds.): America's New Market Geography. Center for Urban Policy Research, New Brunswick, 43-99

Castells M., Hall P. (1994): Technopoles of the World: the Making of 21st Century Industrial Complexes. Routledge, London

CEC (1991): Europe 2000: Outlook for the Development of the Community's Territory, Commission of the European Communities. Directorate-General for Regional Policy, Bruxelles

Chinitz B. (1961): Contrasts in Agglomeration: New York and Pittsburgh. American Economic Review Papers and Proceedings, 279-289

Chow G. (1967): Technological Change and The Demand for Computers. American Economic Review, 57, 1117-1130

Christaller W. (1933): Die Zentralen Orte in Suddeutschland. Fischer, Jena

Ciciotti E. (1992): Le prestazioni innovative delle regioni italiane. In: Boitani A., Ciciotti E. (Eds.): Innovazione e competitività nell'industria italiana. Il Mulino, Bologna, 99-164

Ciciotti E. (1993): Competitività e territorio. Nuova Italia Scientifica, Roma

Coleman W.D., Jacek H.J. (1989): Regionalism, Business Interests and Public Policy. Sage, London

Conti S., Malecki E. J., Oinas P. (1995): The Industrial Enterprise and Its Environment: Spatial Perspectives. Ashgate, Aldershot

Coombs R., Saviotti P., Walsh V. (1987): Economics and Technological Change. McMillan, Basingstoke

Crouch C., Streeck W. (Eds.) (1996): Les capitalismes en Europe. La Découverte, Paris

D'Aspremont C., Gabszewicz J., Thisse J.-F. (1979): On Hotelling's Stability in Competition. Econometrica, 47, 1145-1150

David P.A. (1969): A Contribution to the Theory of Diffusion, Centre for Research in Economic Growth Research Memorandum, 71, Stanford University, Stanford

David P.A. (1984): The Marshallian Dynamics of Industrial Localization: Chicago 1850-1890. Paper presented at the Meeting of the Economic History Association, Chicago, September 21-23

David P.A., Rosenbloom J.L. (1990): Marshallian Factor, Market Externalities and the Dynamics of Industrial Localisation. Journal of Urban Economics, 28, 349-370

Davies S. (1979): The Diffusion of Process Innovations. Cambridge University Press, Cambridge

Davis D.R., Weinstein D.E. (1996): Does Economic Geography Matter for International Specialization?. NBER Working Paper Series, 5706, NBER, Cambridge (Mass.)

Dean R.D., McKee D.L. (1970): Regional Economics, Theory and Practice. Free Press, New York

Delors J. (1993): White Paper on Growth, Competitiveness, and Employment. The Challenges and Ways Forward into the 21st Century. The European Commission, Brussels

Dendrinos D.S. (1996): Cities as Spatial Chaotic Attractors. In: Kiel D.G., Elliot E. (Eds.): Chaos Theory, in the Social Sciences: Foundations and Applications. University of Michigan Press, Ann Arbour, 237-279

Dendrinos D.S.- Mullally H. (1985): Urban Evolution. Studies in the Mathematical Ecology of Cities, Oxford University Press, Oxford

Dendrinos D.S., Rosser J.B. (1992): Fundamental Issues in Nonlinear Urban Population Dynamic Models. Theory and a Synthesis. Annals of Regional Science, 26, 135-145

Dendrinos D.S., Sonis M. (1986): Variational Principles and Conservation Conditions in Volterra's Ecology and in Urban Relative Dynamics. Journal of Regional Science, 26 (2), 359-377

De Smidt M., Wever E. (Eds.) (1990): The Corporate Firm in a Changing World Economy: Case Studies in the Geography of Enterprise. Routledge, London

Dixit A., Stiglitz J. (1977): Monopolistic Competition and Optimum Product Diversity. American Economic Review, 67, 297-308

Dollar D., Wolff E.N. (1993): Competitiveness, Convergence and International Specialisation. MIT Press, Cambridge (Mass.)

Dorfman N.S. (1983): Route 128: the Development of a Regional High Technology Economy. Research Policy, 12, 299-316

Dosi G., Freeman C., Nelson R., Silverberg G., Soete L. (Eds.) (1988): Technical Change and Economic Theory. Pinter, London

Doz Y. (1987): International Industries: Fragmentation versus Globalisation. In: Guile B., Brooks H. (Eds.): Technology and Global Industry. National Academic Press, New York, 96-118

Dumais G., Ellison G., Glaeser E. (1997): Geographic Concentration as a Dynamic Process. NBER Working Paper Series, 6270, NBER, Cambridge (Mass.)

Eaton B.C., Lipsey R.G. (1976): The Introduction of Space into the Neoclassical Model of Value Theory. In: Greenhut M.L., Norman G. (Eds.): The Economics of Location. Elgar, London, 69-106

Edquist C. (1995): The Role of Governments and Markets in Science and Technology Innovation Policies. Paper presented at the workshop "Role and Impact of Science and

Technology in Innovation and Regional Economic Development". University of British Columbia, Victoria, August 23-25
Eichengreen B. (1993): Labour Market and European Monetary Reunification. In: Masson P.R., Taylor M.P. (Eds.): Policy Issues in the Operation of Currency Unions. Cambridge University Press, Cambridge, 130-162
Ellison G., Glaeser E.L. (1997): Geographic Concentration in US Manufacturing Industries: a Dartboard Approach. Journal of Political Economy, 105 (5), 889-927
Farrell J., Saloner G. (1986a): Standardization and Variety. Economic Letters, 20, 71-83
Farrell J., Saloner G. (1986b): Competition, Compatibility and Standards: The Economics of Horses, Penguins, and Lemmings. In: Landis Gabel H. (Ed.): Product Standardization and Competitive Strategy. Elsevier, Amsterdam, 1-21
Feldman M.P. (1994): The University and Economic Development: The Case of Johns Hopkins University and Baltimore. Economic Development Quarterly, 8 (1), 67-76
Feldman M.P. (1995): The Geography of Innovation. Kluwer, Dordrecht
Feldman M.P., Florida R. (1994): The Geographic Sources of Innovation: Technological Infrastructure and Product Innovation in the United States. Annals of American Geographer, 84 (2), 210-229
Fetter F.A. (1924): The Economic Law of Market Areas. Quarterly Journal of Economics, 39, 520-529
Florence P.S. (1948): Investment, Location and Size of Plant. Cambridge University Press, Cambridge
Folloni G., Maggioni M.A. (1994): Un modello dinamico di crescita regionale: leader e attività indotte. Quaderni della ricerca di base: Modelli di sviluppo e regional competition, 3, Università Bocconi, Milano
Freeman C. (1982): The Economics of Industrial Innovation. Pinter, London
Fudenberg D., Tyrole J. (1985): Preemption and Rent Equalisation in the Adoption of New Technology. Review of Economic Studies, 52, 383-401
Fujita M. (1985): Towards General Equilibrium Models of Urban Land Use. La Revue Economique, 36, 135-167
Fujita M. (1986): Optimal Location of Public Facilities: Area Dominance Approach. Regional Science and Urban Economics, 16, 241-268
Fujita M. (1988): A Monopolistic Competition Model of Spatial Agglomeration: Differentiated Product Approach. Regional Science and Urban Economics, 18, 87-124
Fujita M. (1990): Spatial Interactions and Agglomeration in Urban Economics. In: Chatterji M., Kuenne R.E. (Eds.): New Frontiers of Regional Science. McMillan, London, 184 -221
Fujita M., Smith T.E. (1990): Additive Interaction Model of Spatial Agglomeration. Journal of Regional Science, 30 (1), 51-74
Fujita M., Thisse J.-F. (1996): Economics of Agglomeration. CEPR Discussion Paper Series, 1344, CEPR, London
Gabszewicz J.J., Thisse J.-F. (1986): Spatial Competition and the Location of Firms. Fundamentals of Pure and Applied Economics, 5, 305-316
Gabszewicz J.J., Thisse J.-F. (1992): Location. In: Aumann R.J., Hart S. (Eds.): Handbook of Game Theory. Elsevier, Amsterdam, 282-304
Gambarotto F., Maggioni M.A. (1998): Regional Development Strategies in Changing Environments: an Ecological Approach. Regional Studies, 32, 49-61
Garofoli G. (1991): Modelli locali di sviluppo. Angeli, Milano
Gause G.F. (1934): The Struggle for Existence. Williams & Wilkins, Baltimore

Geneau de Lamarlière I. (1991): The Determinants of the Location of the Semiconductor Industry. In: Benko G., Dunford M. (Eds.): Industrial Change and Regional Development: The Transformation of New Industrial Spaces. Pinter, London, 171-189

Geroski P.A. (1991): Market Dynamics and Entry. Blackwell, Oxford

Geroski P.A. (1993): Innovation, Technological Opportunity, and Market Structure. In: Mansfield E. (Ed.): The Economics of Technical Change. Elgar London, 161-177

Geroski P.A. (1995): What Do We Know About Entry?. International Journal of Industrial Organization, 13 (4), 421-440

Geroski P.A., Small I., Walters C.F. (1998): Agglomeration Economies, Technology Spillovers and Company Productivity Growth. CEPR Discussion Paper Series, 1867, CEPR, London

Gibb J.M. (1985): Science Parks and Innovation Centres: Their Economic and Social Impact. Elsevier, Amsterdam

Gibson D.V., Kometsky, G., Smilor R.W. (1992): The Technopolis Phenomenon: Smart Cities, Fast Systems, Global Networks. Rowman, Lanham

Glaeser E.L., Kallal H.D., Scheinkman J.A., Shleifer A. (1992): Growth in Cities. Journal of Political Economy, vol. 100, n. 6, 1126-52

Glaister S. (1972): Advertising Policy and Returns to Scale in Markets where Information Is Passed Between Individuals. Economica, 41, 139-156

Goldstein G.S., Gronberg T.J. (1984): Economies of Scope and Economies of Agglomeration. Journal of Urban Economics, 16, 91-104

Greenaway D., Torstensson J. (1998): Economic Geography, Comparative Advantage and Trade within Industries; Evidence from the OECD. CEPR Discussion Paper Series, 1857, CEPR, London

Greenhut M.L. (1956): Plant Location and Practice. University of North Carolina Press, Chapel Hill

Greenhut M.L., Norman G. (Eds.) (1995): The Economics of Location. Elgar, London

Grilliches Z. (1957): Hybrid Corn: an Exploration in the Economics of Technological Change. Econometrica, 25 (4), 501-522

Hakansson H. (1987) (Ed.), Industrial Technological Development: A Network Approach. Croom Helm, London

Hakansson H. (1989): Corporate Technolgical Behaviour: Co-operation and Networks. Routledge, London

Hall R.E. (1991): Booms and Recessions in a Noisy Economy. Yale University Press, New York

Hall P., Markusen A.R. (Eds.) (1985): Silicon Landscapes. Allen and Unwin, Boston

Hannan M.T., Carrol G. (1992): Dynamics of Organizational Populations: Density, Legitimation, and Competition. Oxford University Press, Oxford

Hannan M.T., Freeman J. (1989): Organizational Ecology. Harvard University Press, Cambridge (Mass.)

Hanson G. H. (1996): Agglomeration, Dispersion and the Pioneer Firm. Journal of Urban Economics, 39, 255-281

Harrington J.W., Warf B. (1995): Industrial Location. Principles, Practice and Policy. Routledge, London

Henderson J.V. (1977): Economic Theory and the Cities. National Academic Press, New York

Henderson J.V. (1994): Externalities and Industrial Development. NBER Working Paper Series, 4730, NBER, Cambridge (Mass.)

Hilpert U. (Ed.) (1991): Regional Innovation and Decentralization: High Tech Industry and Government Policy. Routledge, London

Hirshleifer D. (1993): The Blind Leading the Blind: Social Influences, Fads, and Informational Cascades. Working Paper of the Anderson Graduate School of Management, 24. UCLA, Los Angeles

Hirshleifer D., Welch I. (1994): Institutional Memory, Inertia and Impulsiveness. Working Paper of the Anderson Graduate School of Management, 2. UCLA, Los Angeles

Hirschman A.O. (1958): The Strategy of Economic Development. Yale University Press, New Haven

Hofbauer J., Sigmund K. (1984): The Theory of Evolution and Dynamical Systems. Cambridge University Press, Cambridge

Hoover E.M. (1937): Spatial Price Discrimination. Review of Economic Studies, 4, 182-191

Hoover E.M. (1948): The Location of Economic Activities. McGraw-Hill, New York

Hoover E.M. (1971): An Introduction to Regional Economics. Knopf, New York

Hotelling H. (1929): Stability in Competition. Economic Journal, 39, 41-57

Howells J.R.L. (1984): The Location of Research and Development: Some Observations and Evidence from Britain. Regional Studies, 18 (1), 13-29

Ireland N., Stoneman P. (1986): Technological Diffusion, Expectations and Welfare. Oxford Economic Papers, 38, 283-304

Isard W. (1949): The General Theory of Location and Space-Economy. Quarterly Journal of Economics, LXIII, 476-506

Isard W. (1951): Distance Inputs and the Space Economy: the Locational Equilibrium of the Firm. Quarterly Journal of Economics, 65, 373-397

Isard W. (1956): Location and Space-economy. MIT Press, Cambridge (Mass.)

Isard W. (1960): Methods of Regional Analysis: An Introduction to Regional Science. Wiley, New York

Isard W. (1966): Game Theory, Location Theory, and Industrial Agglomeration. Papers of the Regional Science Association, 14, 1-33

Isard W. (1969): General Theory: Social, Political, Economic and Regional. MIT Press, Cambridge (Mass.)

Isard W.- Schooler E.W. (1959): Industrial Complex Analysis, Agglomeration Economies and Regional Development. Journal of Regional Science, 1, 19-34

Isard W., Smith C. (1967): Location Games: With Application to Classic Location Problems. Papers of the Regional Science Association, 19, 45-80

Jacobs (1964): The Death and Life of Great American Cities. Penguin and Cape, Harmondsworth

Jacobs (1969): The Economy of Cities. Random House, New York

Jacobs (1984): Cities and the Wealth of Nations: Principles of Economic Life. Random House, New York

Jaffe A.B. (1989): Real effects of Academic Research. American Economic Review, 79 (5), 957-979

Jaffe A.B., Henderson R., Trajtenberg M. (1993): Geographic Localization of Knowledge Spillovers as Evidenced by Patent Citations. Quarterly Journal of Economics, 63 (3), 577-598

Kaldor N. (1970): The Case for Regional Policies. Scottish Journal of Political Economy, (3), 337-348

Kanemoto Y. (1980): Theories of Urban Externalities. North-Holland, Amsterdam

Kanemoto Y. (1990): Optimal Cities with Indivisibility in Production and Interactions between Firms. Journal of Urban Economics, 27, 45-59

Karshenas M., Stoneman P. (1993): Rank, Stock, Order and Epidemic Effects in the Diffusion of New Process Technologies: an Empirical Model. Rand Journal of Economics, 24 (4), 503-528

Karshenas M., Stoneman P. (1995): Technological Diffusion. In: Stoneman P., Handbook of the Economics of Innovation and Technological Change. Blackwell, Oxford, 265-297

Keating M., Loughlin J. (1997): The Political Economy of Regionalism, Case Studies in Regional and Federal Studies. Cass, London

Keeble D.E. (1988): High Technology Industry and Local Environments in the United Kingdom. In: Aydalot P., Keeble D. (Eds.): High Technology Industry and Innovative Environments: the European Experience. GREMI-Routledge, London, 69-98

Keeble D.E., Kelly T. (1986): New Firms and High-Technology in the United Kingdom: The case of Computer Electronics. In: Keeble D., Wever E. (Eds.): New Firms and Regional Development in Europe. Croom Helm, London, 75-104

Keeble D.E., Wever E. (Eds.) (1986): New Firms and Regional Development in Europe. Croom Helm, London

Kim S. (1995): Expansion of Markets and the Geographic Distribution of Economic Activities. Quarterly Journal of Economics, 3, 881-908

Kim S. (1997): Regions, Resources and Economic Geography: Sources of US Regional Comparative Advantage: 1880-1987. NBER Working Paper Series, 6322, NBER, Cambridge (Mass.)

Kingsland S.H. (1985): Modelling Nature. Episodes in the History of Population Ecology. The University of Chicago Press, Chicago

Knoblauch V. (1991): Generalizing Location Games to a Graph. The Journal of Industrial Economics, 39, 683-687

Krugman P. (1990): Rethinking International Trade. MIT Press, Cambridge (Mass.)

Krugman P. (1991a): Geography and Trade. MIT Press, Cambridge (Mass.)

Krugman P. (1991b): Increasing Returns and Economic Geography. Journal of Political Economy, 99 (3), 483-499

Krugman P. (1991c): History vs. Expectations. Quarterly Journal of Economics, 106 (2), 651-667

Krugman P. (1992): A Dynamic Spatial Model. NBER Working Paper Series, 4219, NBER, Cambridge (Mass.)

Krugman P. (1993a): On the Number and Location of Cities. European Economic Review, 37, 129-144

Krugman P. (1993b): First Nature, Second Nature, and Metropolitan Location. Journal of Regional Science, 33, 129-144

Krugman P. (1993c): On the Relationship between Trade Theory and Location Theory. Review of International Economics, 1 (2), 110-122

Krugman P. (1994): Complex Landscapes in Economic Geography. American Economic Review, 84 (2), 412-416

Krugman P. (1995): Development, Geography and Economic Theory. MIT Press, Cambridge (Mass.)

Krugman P. (1998): Space: The Final Frontier. Journal of Economic Perspectives, 12 (2), 161-174

Levin S.A. (1970): Community Equilibria and Stability, and an Extension of the Competitive Exclusion Principle. American Naturalist, 108, 207-228
Levins R. (1968): Evolution in Changing Environments: Some Theoretical Explorations. Princeton University Press, Princeton
Linda R. (1986): Competition Policies and Measures of Dominant Power. In: de Jong H. W., Sheperd W.G. (Eds.): Mainstreams in Industrial Organization. Kluwer, Dordrecht, 287-307
Longhi C., Queré M. (1991): La technopôle comme système industriel localisé: éléments d'analyse et d'enseignements empiriques. Economies et Sociétés, 25 (8), 21-41
Lösch A. (1954): The Economics of Location. Yale University Press, New Haven
Lotka A.J. (1925): Elements of Physical Biology. Williams & Wilkins, Baltimore
Luger M.I., Goldstein H.A. (1991): Technology in the Garden: Research Parks and Regional Economic Development. The University of North Carolina Press, Chapel Hill
Lund L. (1986): Locating Corporate R&D Facilities. Research Report, n. 892. The Conference Board, New York
Macdonald S. (1986): Headhunting in High Technology. Technovation, 4, 233-245
Maggioni M.A. (1990): Sostegno reale all'innovazione: lo strumento degli Science Parks. Un confronto internazionale per l'analisi della realtà italiana. Dynamis-Quaderni IDSE, 10/90, IDSE-CNR, Milano
Maggioni M.A. (1993): Ecological Dynamics and Critical Mass Processes in the Location of High Tech Firms. Paper presented at the 40th RSAI (Regional Sciences Association International) Conference, North American Section, Houston, 11-14 November
Maggioni M.A. (1994): Modelli ecologici per l'analisi della dinamica industriale regionale. In: Pasquini F., Pompili T., Secondini P. (Eds.): Modelli d'analisi e d'intervento per un nuovo regionalismo. Angeli, Milano, 79-105
Maggioni M.A. (1995): The Economic Analysis of Science and Technology Parks: Theoretical Suggestions and the Italian Experience. Sviluppo Economico, 1 (1-2), 149-177
Maggioni M.A. (1999): Clustering Dynamics and the Location of High-tech Firms. Ph.D. Thesis, Warwick Business School, Coventry.
Maggioni M.A., Gambarotto F. (1997): Sviluppo locale, dinamiche globali e ruolo dell'operatore pubblico: un modello ecologico. In: Bramanti A., Maggioni M.A. (Eds.): La dinamica dei sistemi produttivi territoriali: teorie, tecniche, politiche. Angeli, Milano, 280-306
Maggioni M.A., Miglierina C. (1995): Dov'è il motore del sistema tecnologico nazionale? Un'analisi spaziale dei flussi innovativi intersettoriali. In: Gorla G., Vito Colonna O. (Eds.): Regioni e sviluppo: modelli, politiche e riforme. Angeli, Milano, 79-114
Maggioni M.A., Porro G. (1994): Dinamiche di crescita regionale: il ruolo delle aspettative in modelli di tipo ecologico. Quaderni della ricerca di base: Modelli di sviluppo e regional competition, 6, Università Bocconi, Milano
Mahajan V., Peterson R.A. (1985): Models for Innovation Diffusion. Sage, Newbury Park
Mahajan V., Schoeman M.E.F. (1977): Generalized Model for the Time Pattern of the Diffusion Process. IEEE Transactions on Engineering Management, 24, 12-18
Maino R. (1989): Aree interne ed interdipendenze settoriali. Uno schema interpretativo fondato su modelli dinamici del tipo preda-predatore. In: Becchi Collidà A., Ciciotti E., Mela A. (Eds.): Aree interne, tutela del territorio e valorizzazione delle risorse. Angeli, Milano, 81-110
Malecki E.J. (1979): Locational Trends in R&D by Large US Corporation: 1965-1977. Economic Geography, 55, 309-323

Malecki E.J. (1983): Technology and Regional Development: a Survey. International Regional Science Review, 8 (2), 89-125

Malecki E.J. (1986): Research and Development and the Geography of High Technology Complexes. In: Rees J. (Ed.): Technology, Regions and Policy. Rowman, Totowa, 51-74

Malecki E.J. (1987): The R&D Location Decision of the Firm and "Creative" Regions. Technovation, 6, 205-222

Malecki E.J. (1991): Technology and Economic Development: the Dynamics of Local, Regional and National Change. Longman, Harlow

Malecki E.J., Varaiya P. (1986): Innovation and Changes in Regional Structure. In: Nijkamp P. (Ed.): Handbook of Regional and Urban Economics, 1. Elsevier, Amsterdam, 629-645

Mansfield E. (1968): Industrial Research and Technological Innovation. Norton, New York

Markusen A.R. (1987): Regions: the Economics and Politics of Territory. Rowman and Allanheld, Totowa

Markusen A.R., Hall P., Glassmeier A. (1986): High-tech America. Allen and Unwin, Boston

Marshall A. (1920): Principles of Economics. McMillan, London

Marshall A . (1921): Industry and Trade. McMillan, London

Marshall M. (1987): Long Waves of Regional Development. McMillan, London

Maskell P. (1997): Apprendimento localizzato e competitività industriale. In: Bramanti A., Maggioni M.A. (Eds.): La dinamica dei sistemi produttivi territoriali: teorie, tecniche, politiche. Angeli, Milano, 112-133

Maskell P., Heskelinen H., Hannibalsson I., Malmberg A., Vatne E. (1998): Competitiveness, Localised Learning and Regional Development: Specialisation and Prosperity in Small Open Economies. Routledge, London

May R.M. (1974): Stability and Complexity in Model Ecosystems. Princeton University Press, Princeton

May R.M. (1976): Theoretical Ecology, Principles and Applications. Saunders, Philadelphia

Metcalfe J.S. (1994): Competition, Fisher's Principle and Increasing Returns in the Selection Process. Journal of Evolutionary Economics, 4 (4), 327-346

Meyer D.R. (1963): Regional Economics: a Survey. American Economic Review, 53, 19-54

Miyao T., Kanemoto Y. (1987): Urban Dynamics and Urban Externalities. Harwood Academic Publishers, London

Molle W., Beumer L., Boeckhout I. (1989): The Location of Information Intensive Activities in the European Community. In: Punset E., Sweeney G. (Eds.): Information Resources and Corporate Growth. Pinter, London, 161-172

Monck C.S.P., Porter R.B., Quintas P., Storey D.J., Wynarczyk P. (1988): Science Parks and the Growth of High Technology Firms. Croom Helm, London

Moriarty B.M. (1980): Industrial Location and Community Development. University of North Carolina Press, Chapel Hill

Morita K., Hiraoka H. (1988): Technopolis Osaka: Integrating Urban Functions and Science. In: Smilor R.W., Kozmetsky G., Gibson D.V. (Eds.): Creating the Technopolys: Linking Technology, Commercialization and Economic Development. Ballinger, Cambridge (Mass.), 23-50

Moses L.N. (1958): Location and the Theory of Production. Quarterly Journal of Economics, 72, 259-272

Murray G.C. (1994): The European Union's Support for New Technology-based Firms: An Assessment of the First Three Years of the European Seed Capital Fund Scheme. European Planning Studies, 2 (4), 435-461

Murray G.C. (1996): A Policy Response to Regional Disparities in the Supply of Risk Capital to New Technology-based Firms in the European Union: The European Seed Capital Fund Scheme. Regional Studies, 32 (5), 405-419

Murray G.C., Lott J.L. (1995): Have UK Venture Capitalist a Bias Against Investment in New Technology-based Firms?. Research Policy, 24, 283-299

Nijkamp P. (Ed.) (1986): Handbook of Regional and Urban Economics. Elsevier, Amsterdam

Nijkamp P., Reggiani A. (1990): Theory of Chaos: Relevance for Analysing Spatial Processes. In: Fisher M.M., Nijkamp P., Papageorgiou Y.Y. (Eds.): Spatial Choices and Processes. Elsevier, Amsterdam, 49-79

Nijkamp P., Reggiani A. (1991): Chaos Theory and Spatial Dynamics. Journal of Transport Economics and Policy, 25 (1), 81-96

Nijkamp P., Reggiani A. (1992): Interaction, Evolution and Chaos in Space. Springer-Verlag, Berlin

Nijkamp P., Reggiani A. (1998): The Economics of Complex Spatial Systems. Elsevier, Amsterdam

North D.C. (1955): Location Theory and Regional Economic Growth. Journal of Political Economy, (3), 243-258

Norton R.D. (1992): Agglomeration and Competitiveness: From Marshall to Chinitz. Urban Studies, 29 (2), 155-170

Oakey R. (1981a): High Technology Industry and Industrial Location. Gower, Farnbourough

Oakey R. (1981b): High Technology Industry and Industrial Development. In: Hamilton F.E., Linge G.J. (Eds.): Spatial Analysis. Industry and the Industrial Environment, 279-295

Oakey R.P. (1984): High Technology Small Firms. Regional Development in Britain and the United States. Pinter, London

Oakey R.P. (Ed.) (1994): New Technology-Based Firms in the 1990s. Chapman, London

Oakey R.P., Thwaites A.T., Nash P.A. (1980): The Regional Distribution of Innovative Manufacturing Establishments in Britain. Regional Studies, 14, 235-253

OECD (1986): Main Science and Technology Indicators. OECD, Paris

OECD (1997): Main Science and Technology Indicators. OECD, Paris

OTA (1984): Technology, Innovation, and Regional Economic Development. OTA, Washington

Ottaviano G.I.P., Puga P. (1997): Agglomeration in The Global Economy: a Survey of the "New Economic Geography". CEPR Discussion Paper Series, 1699, CEPR, London

Paci R., Usai S. (1997) Technological enclaves and industrial districts. An Analysis of the Regional Distributions of Innovative Activity in Europe. CRENOS Working Paper, 97/8, Università di Cagliari, Cagliari

Padmore T., Gibson H. (1998): Modelling Systems of Innovation: II. A framework for Industrial Cluster Analysis in Regions. Research Policy, 26, 625-41

Palander T. (1935): Beitrage zur Standortstheorie. Almqvist und Wicksells, Uppsala

Papageorgiou Y.Y. (1979): Agglomerations. Regional Science and Urban Economics, 9, 41-49

Papageorgiou Y.Y., Thisse J.-F. (1985): Agglomeration as Spatial Interdependence Between Firms and Households. Journal of Economic Theory, 37, 19-31

Papke L.E. (1991): Interstate Business Tax Differentials and New Firm Location. Evidence from Panel Data. Journal of Public Economics, 45, 47-68

Pascal A.H., McCall J.J. (1980): Agglomeration Economies, Search Costs and Industrial Location. Journal of Urban Economics, 8, 383-388

Pavitt K. (1984): Sectoral Patterns of Technical Change: towards a Taxonomy and a Theory. Research Policy, 6, 343-372

Pearl R., Reed L.J. (1925): Skew-Growth Curves. Proceeding of the National Academy of Natural Sciences of the USA, 11, 16-22

Perroux F. (1950): Economic Space: Theory and Applications. Quarterly Journal of Economics, 64 (1), 89-104

Perroux F. (1955): Note sur la notion de pôle de croissance. Economie Appliquée, 8, 307-320

Ponsard C. (1983): History of Spatial Economic Theory. Springer, Berlin

Porter M. (1980): Competitive Strategy. Free Press, New York

Porter M. (1985): Competitive Advantage. Free Press, New York

Porter M. (1990): The Competitive Advantage of Nations. Free Press, New York

Porter M. (1996): Competitive Advantage, Agglomeration Economies, and Regional policy. International Regional Science Review, 19 (1-2), 85-94

Preer R.W. (1992): The Emergence of Technopolis, Knowledge-intensive Technologies and Regional Development. Praeger, New York

Premus R. (1982): Location of High-Technology Firms and Regional Economic Development: a Staff Study, Joint Economic Committee, Congress of the United States. US Government Printing Office, Washington

Premus R. (1984): Venture Capital and Innovation. A Study Prepared for the Joint Economic Committee, Congress of the United States. US Government Printing Office, Washington

Puga D., Venables A.J. (1996): The Spread of Industry: Spatial Agglomeration in Economic Development. CEPR Discussion Paper Series, 1354, CEPR, London

Puga D., Venables A.J. (1998): Agglomeration and Economic Development: Import Substitution versus Trade Liberalisation. CEPR Discussion Paper Series, 1782, CEPR, London

Quirmbach H.C. (1986): The Diffusion of New Technology and The Market for an Innovation. Rand Journal of Economics, 17, 33-47

Ratti R., Bramanti A., Gordon R. (1997): The Dynamic of Innovative Regions: the GREMI Approach. Avebury, Aldershot

Rees J., Stafford H. (1983): Theories of Regional Growth and Industrial Location: Their Relevance for High Technology Industry in the United States. OTA, Washington

Reggiani A. (1997): Towards the Complex City: Approaches and Experiments. In: Bertuglia C.S., Bianchi G., Mela A. (Eds.): The City and Its Sciences. Springer, Berlin

Regional Studies vol. 28, (4), January 1994, Special Issue on New Firms Formation

Reinganum J.F. (1981): On the Diffusion of New Technology: A Game Theoretic Approach. Review of Economic Studies, XLVIII, 395-405

Richardson H. W. (1978): Regional and Urban Economics. Penguin Books, Harmondsworth

Rivera-Batiz F.L. (1988): Increasing Returns, Monopolistic Competition and Agglomeration Economies in Consumption and Production. Regional Science and Urban Economics, 18, 125-153

Rombaldoni R., Zazzaro A. (1997): Localizzazione delle imprese manifatturiere e specializzazione regionale in Italia: 1970-1990. Rassegna di lavori dell'ISCO, XIV (1), 151-181

Romer P. (1986): Increasing Returns and Long-run Growth. Journal of Political Economy, 94 (5), 1002-1037

Roughgarden J. (1979): Theory of Population Genetics and Evolutionary Ecology: an Introduction. McMillan, New York-London

Rowe D. (1988): Science Parks: The UK Experience. In: Campodall'Orto S., Roveda C., Parchi scientifici come strumenti di politica industriale. Angeli, Milano, 35-57

Salais R., Storper M. (1992): The Four "Worlds" of Contemporary Industry. Cambridge Journal of Economics, 16, 169-193

Salais R., Storper M. (1993): Les mondes de production. Enquête sur l'identité économique de la France. Éditions de l'École des Hautes Études en Sciences Sociales, Paris

Saxenian A. (1994): Regional Advantage, Culture and Competition in Silicon Valley and Route 128. Harvard University Press, Cambridge (Mass.)

Schelling T.C. (1978): Micromotives and Macrobehaviour. Norton, New York

Scherer F.M. (1982): Inter-Industry Technology Flows in the United States. Research Policy, 11, 227-245

Schmenner R.W. (1977): Urban Industrial Location: an Evolutionary Model. Journal of Regional Science, 17 (2), 179-94

Schmenner R.W. (1982): Making Business Location Decision. Prentice Hall, Englewood Cliffs

Scotchmer S., Thisse J.-F. (1992): Space and Competition. The Annals of Regional Science, 26, 269-286

Scott J. (1986): Industrial Organization and Location: Division of Labour, the Firm, and Spatial Process. Economic Geography, 62 (3), 215-231

Scott A.J. (1988a): New Industrial Spaces: Flexible Production and Regional Development in North America and Western Europe. Pion, London

Scott A.J. (1988b): Metropolis. From the Division of Labour to Urban Form. University of California Press, Los Angeles

Scott A.J. (1993): Technopolis. High-Technology Industry and Regional Development in Southern California. University of California Press, Berkley

Scott A.J., Storper M. (1992): Regional Development Reconsidered. In: Ernst H., Meier V. (Eds.): Regional Development and Contemporary Industrial Response: Extending Flexible Specialisation. Belhaven, London, 61-86

Smilor R.W., Kozmetsky G., Gibson D.V. (Eds.) (1988): Creating the Technopolys: Linking Technology, Commercialization and Economic Development. Ballinger, Cambridge (Mass.)

Sonis M. (1986): A Unified Theory of Innovation Diffusion, Dynamic Choice of Alternatives, Ecological Dynamics and Urban/Regional Growth and Decline. In: Ricerche Economiche, 4, 696-723

Staber U. (1997): An Ecological Perspective on Entrepreneurship in Industrial Districts. Entrepreneurship and Regional Development, 9, 45-64

Stahl K. (1987): Theories of Urban Business Location. In: Mills E.S. (Ed.): Handbook of Regional and Urban Economics. North Holland, Amsterdam, 759-820

Starret D. (1978): Market Allocation of Location Choice in a Model with Free Mobility. Journal of Economic Theory, 17, 21-37

Sternberg R. (1997): Does Location Matters? On the Impact of Innovation Centres on the Development of Innovation Oriented Start-ups. Paper presented at the workshop "Firms Dynamics in High-Technology Industries", ZEW, Mannheim, 9-10 June

Stoneman P. (1983): The Economics Analysis of Technological Change. Oxford University Press, Oxford

Stoneman P. (1986): Technological Diffusion: The Viewpoint of Economic Theory. Ricerche Economiche, 4, 585-606

Stoneman P. (1995): Handbook of the Economics of Innovation and Technological Change. Blackwell, Oxford

Storey D.J. (1982): Entrepreneurship and the New Firm. Croom Helm, London

Storey D.J. (1985): Small Firms in Regional Economic Development: Britain, Ireland, and the United States. Cambridge University Press, Cambridge

Storey D.J., Tether B.S. (1998a): New Technology-Based Firms in the European Union: an Introduction. Research Policy, 26, 933-946

Storey D.J., Tether B.S. (1998b): Public Policy Measures to Support New Technology-Based Firms in the European Union. Research Policy, 26, 1307-1357

Storper M. (1997): The Regional Word: Territorial Development in a Global Economy. Guilford Press, New York

Storper M., Scott A.J. (1989): The Geographical Foundation and Social Regulation of Flexible Production Complexes. In: Wolch J., Dear M. (Eds.): The Power of Geography: how Territory Shapes Social Life. Unwin, Boston, 21-40

Storper M., Walker R. (1984): The Spatial Division of Labour: Labour and the Location of Industries. In: Sawers L., Tabb W.K. (Eds.): Sunbelt/Snowbelt: Urban Development and Regional Restructuring. Oxford University Press, Oxford, 19-47

Storper M., Walker R. (1989): The Capitalist Imperative. Territory, Technology and Industrial Growth. Basil Blackwell, New York

Swann G.M.P. (1993): Can High Technology Services Prosper if High Technology Manufacturing Doesn't?. Centre for Business Strategy Working Paper Series, 143, London Business School, London

Swann G.M.P. (1998): Towards a Model of Clustering in High-Technology Industries. In: Swann G.M.P., Prevezer M., Stout D. (Eds.): The Dynamics of Industrial Clustering. Oxford University Press, Oxford, 52-76

Swann G.M.P., Prevezer M. (1996): A Comparison of the Dynamics of Clustering in Computing and Biotechnology. Research Policy, 25, 1139-1157

Swann G.M.P., Prevezer M., Stout D. (Eds.) (1998): The Dynamics of Industrial Clustering. Oxford University Press, Oxford

Tauchen H., Witte A.D. (1983): An Equilibrium Model of Office Location and Contact Patterns. Environment and Planning A, 15, 1311-1326

Taylor M., Conti S. (1997): Interdependent and Uneven Development: Global-Local Perspectives. Ashgate, Aldershot

Temple P. (1998): Cluster and Competitiveness, A Policy Perspective. In: Swann G.M.P., Prevezer M., Stout D. (Eds.): The Dynamics of Industrial Clustering. Oxford University Press, Oxford, 257-297

Thirtle C G., Ruttan V.W. (1987): The Role of Demand and Supply in the Generation and Diffusion of Technical Change. Harwood Academic, Chur

Tiebout C. (1956): Exports and Regional economic Growth. Journal of Political Economy, 2, 160-169

Townroe P.M., Roberts N.J. (1980): Local External Economies for British Manufacturing Industry. Gower, Farnborough

Twaites A.T., Oakey R.P. (1985): The Regional Economic Impact of Technological Change. Pinter, London

Venables A.J. (1995): Economic Integration and the Location of Firms. American Economic Review, 85 (2), 296-300

Venables A.J. (1996): Equilibrium Locations of Vertically Linked Industries. International Economic Review, 37, 341-359

Verhulst P.F. (1845) Recherches Mathématique sur la loi d'accroissement de la population. Nouveaux Mémoires de l'Académie Royale des Sciences et Belles-Lettres de Bruxelles, 18 (2), 3-38

Vivarelli M. (1994): La nascita delle imprese in Italia. Teorie e verifiche empiriche. Egea, Milano

Vivarelli M. (1995): The Economics of Technology and Employment: Theory and Empirical Evidence. Ashgate, Aldershot

Volterra V. (1926): Variazioni e fluttuazioni del numero di individui in specie di animali conviventi. Memorie Accademia dei Lincei, 2, 31-113

von Hagen J., Hammond G.W. (1998): Regional Insurance Against Asymmetric Shocks: An Empirical Study for the European Community. Manchester School, 66 (3), 331-353

von Hippel E. (1988): The Sources of Innovation. Oxford University Press, Oxford

von Thünen (1875): Der Isoliert Staat in Beziehung auf Landwirtschaft und Nationalokonomie. Puthes, Hamburg

Walras L. (1874): Eléments d'économie politique pure. Paris

Waterson M. (1984): Economic Theory of the Industry. Cambridge University Press, Cambridge

Weber (1929): Theory of the Location of Industry. Chicago University Press, Chicago

Westhead P., Storey D.J. (1994): An Assessment of Firms Located On and Off Science Parks in the United Kingdom. HMSO, London

Declaration

A paper has been produced from section 4.2 and published as Maggioni M.A. (1994), "Modelli ecologici per l'analisi della dinamica industriale regionale", in Pasquini F., Pompili T, Secondini P., (eds.), *Modelli di analisi e di intervento per un nuovo regionalismo*, Franco Angeli, Milano.

A joint paper with Francesca Gambarotto has been produced from section 4.4 and published as Gambarotto F., Maggioni M.A. (1998), Regional Development Strategies in Changing Environments: an Ecological Approach, *Regional Studies*, 32 (1), pp. 49-62.

A paper has been produced from section 7.7 and published as Maggioni M.A. (1995),"The Economic Analysis of Science and Technology Parks: Theoretical Suggestions and Italian Experience", Sviluppo Economico, n. 1/94-95.

Summary

The location of productive activities and the emergence of clustering dynamics has been an important research topic since the early works of Weber (1929) and Marshall (1920 and 1921). This book aims at relating the processes of firms' location decision and the development of high-tech clusters within an encompassing theoretical and empirical framework. The book shows the empirical relevance of the clustering of high-tech sectors and highlights the importance of the issue through the construction and use of an original database on the location of high-tech establishments and employment (at two different geographical levels) in four major industrialised countries. It also contains a critical review of a number of different streams of theoretical and empirical literature which are directly connected, or which have been explicitly put in connection by the author, with the topic of study. In the book we develop a composite modelling framework for analysing firms' location decisions and the growth of high-tech clusters, and we empirically test a number of crucial hypotheses in order to draw some guidelines for economic policy. The models presented in the theoretical chapter derive from two different streams of literature. The first derives from the analysis of population ecology, the second from the theory of innovation diffusion. These modelling frameworks have stressed the existence of a critical mass and a maximum dimension of the cluster and their effects on the early and late phases of development within the "life cycle" of a cluster. They also highlighted the role of rank, stock, order and epidemics effects in the location decision of an individual firm which has to decide whether to locate into a developing cluster. The empirical evidence presented in the book has focused on the crucial elements of the location process by verifying the empirical relevance of different locational factors, has stressed the relative importance of agglomeration versus scale economies in determining the industrial specialisation of an area, and has measured the competitive effects which arise between the development of different clusters and the synergistic effects which are generated within the cluster. Finally the book presents empirical evidence which shows that local competition and industrial specialisation are the key elements for the success of an industrial cluster. A final chapter extracts some crucial policy conclusions on the role of entry versus growth policies, on the different development path that an industrial cluster may follow depending on the excludability condition, presents an original taxonomy of specific policies, applies some of these findings to a brief survey of the phenomenon of science parks and finally produces a series of guidelines for policy makers. The conclusion summarises the results obtained in the book and present a brief agenda for future research.

List of figures

Figure 2.1	Geographical distribution of integrated circuit manufacturing establishments (1982)	9
Figure 2.2	Location of US manufacturing employment (1990)	9
Figure 2.3	Geometric representation of Gini coefficient	14
Figure 2.4	High-tech location quotients for US at FLA level (establishments)	31
Figure 2.5	High-tech location quotients for US at FLA level (employment)	31
Figure 2.6	High-tech location quotients for UK at FLA level (establishments)	32
Figure 2.7	High-tech location quotients for UK at FLA level (employment)	32
Figure 2.8	High-tech location quotients for France at FLA level (establishments)	33
Figure 2.9	High-tech location quotients for France at FLA level (employment)	33
Figure 2.10	High-tech location quotients for Italy at FLA level (establishments)	34
Figure 2.11	High-tech location quotients for Italy at FLA level (employment)	34
Figure 2.12	US Pharmaceuticals spatial core (employment)	45
Figure 2.13	US Industrial process control instruments spatial core (employment)	45
Figure 2.14	US Optical and photo. equipment spatial core (employment)	46
Figure 2.15	US Aerospace spatial core (establishments)	46
Figure 2.16	US Textiles spatial core (employment)	47
Figure 2.17	US Motor vehicles spatial core (employment)	47
Figure 2.18	UK Pharmaceuticals spatial core (establishments)	48
Figure 2.19	UK Computers and office machinery spatial core (establishments)	48

Figure 2.20 UK Electronic components spatial core (employment)............. 49

Figure 2.21 UK Industrial process control instrum. spatial core (establishments)............. 49

Figure 2.22 UK Optical and photo. equipment spatial core (establishments)............. 50

Figure 2.23 UK Aerospace spatial core (employment)............. 50

Figure 2.24 France Computers and office machin. spatial core (establishments)............. 51

Figure 2.25 France Aerospace spatial core (employment)............. 51

Figure 2.26 France Motor Vehicles spatial core (employment)............. 52

Figure 2.27 Italy Electronic components spatial core (establishments)............. 53

Figure 2.28 Italy Medical and surgical Instruments spatial core (establishments)............. 53

Figure 2.29 Italy Optical and photo. equipment spatial core (employment)............. 54

Figure 2.30 Italy Aerospace spatial core (establishments)............. 54

Figure 4.1 The development path of an industrial cluster............. 100

Figure 4.2 Agglomeration costs and benefits for incumbents and critical sizes of a cluster............. 104

Figure 4.3 Inter-regional competition............. 108

Figure 4.4 Inter-industry facultative mutualism............. 113

Figure 4.5 Core-periphery dynamics (inter-industry commensalism)............. 115

Figure 4.6 Expected locational net benefits for type-*x* firm............. 119

Figure 4.7 Expected locational net benefits for different types of firm............. 121

Figure 4.8 A distribution function of firm's types............. 122

Figure 4.9 Growth of the cluster in two different macro-economic envirnoments............. 132

Figure 4.10 Optimal long run development policies: a graphical representation............. 135

Figure 4.11 Location costs structure and locational gross benefits............. 140

Figure 4.12 Psychological distribution function of potential entrants............. 142

Figure 4.13 The emergence of a critical mass in the development of the cluster............. 143

Figure 7.1 Agglomeration costs and benefits and critical sizes of a cluster............. 208

List of tables

Table 2.1 The original database ... 24
Table 2.2 Location quotients for US at FLA level (establishments) ... 25
Table 2.3 Location quotients for UK at FLA level (establishments) ... 27
Table 2.4 Location quotients for France at FLA level (establishments) ... 28
Table 2.5 Location quotients for Italy at FLA level (establishments) ... 30
Table 2.6 US spatial concentration and inequality indexes at FLA level (establishments) ... 35
Table 2.7 US spatial concentration and inequality indexes at FLA level (employment) ... 36
Table 2.8 UK spatial concentration and inequality indexes at FLA level (establishments) ... 38
Table 2.9 UK spatial concentration and inequality indexes at SLA level (establishments) ... 38
Table 2.10 France spatial concentration and inequality indexes at FLA level (establishments) ... 40
Table 2.11 France spatial concentration and inequality indexes at FLA level (employment) ... 40
Table 2.12 France spatial concentration and inequality indexes at SLA level (establishments) ... 41
Table 2.13 France spatial concentration and inequality indexes at SLA level (employment) ... 41
Table 2.14 Italy spatial concentration and inequality indexes at FLA level (establishments) ... 42
Table 2.15 Italy spatial concentration and inequality indexes at FLA level (employment) ... 43
Table 2.16 Italy spatial concentration and inequality indexes at SLA level (establishments) ... 43
Table 2.17 Italy spatial concentration and inequality indexes at SLA level (employment) ... 44
Table 2.18 High-tech specialisation (total high-tech/total manufacturing) ... 55

Table 2.19 Average size of high-tech establishments (emp./estab.) 55
Table 2.20 Spatial association between high-tech industries in four OECD Countries at FLA level (establishments) 57
Table 3.1 Summary of theoretical approaches 90
Table 4.1 Different types of bilateral interactions.................................... 111
Table 4.2 Optimal long run development policies: a matrix 135
Table 6.1 Geographical factors explaining the location decision of US high-tech firms (cross-section analyses for 1986 and 1993)..... 169
Table 6.2 Geographical factors and the location of US high-tech firms (panel data estimation) 171
Table 6.3 Agglomeration versus scale economies at FLA level................ 173
Table 6.4 Agglomeration versus scale economies at SLA level 174
Table 6.5 Scale versus agglomeration economies in the location of high-tech firms 178
Table 6.6 US high-tech cluster development and inter-state competition coefficients 182
Table 6.7 The development of clusters: inter-regional and inter-industries effects.. 184
Table 6.8 Fastest and slowest growing state-industries (employment) in the period 1975-1995 191
Table 6.9 The engines of high-tech clusters' growth................................. 192
Table 7.1 A taxonomy of goods 209

Contributions to Economics

Frank Bohn
Monetary Union and Fiscal Stability
2000. ISBN 3-7908-1266-8

Jaime Behar
Cooperation and Competition in a Common Market
2000. ISBN 3-7908-1280-3

Michael Malakellis
Integrated Macro-Micro-Modelling Under Rational Expectations
2000. ISBN 3-7908-1274-9

Stefan Baumgärtner
Ambivalent Joint Production and the Natural Environment
2000. ISBN 3-7908-1290-0

Henri Capron, Wim Meeusen (Eds.)
The National Innovation System of Belgium
2000. ISBN 3-7908-1308-7

Tobias Miarka
Financial Intermediation and Deregulation
2000. ISBN 3-7908-1307-9

Chisato Yoshida
Illegal Immigration and Economic Welfare
2000. ISBN 3-7908-1315-X

Nikolaus Thumm
Intellectual Property Rights
2000. ISBN 3-7908-1329-X

Max Keilbach
Spatial Knowledge Spillovers and the Dynamics of Agglomeration and Regional Growth
2000. ISBN 3-7908-1321-4

Alexander Karmann (Ed.)
Financial Structure and Stability
2000. ISBN 3-7908-1332-X

Joos P.A. van Vugt/Jan M. Peet (Eds.)
Social Security and Solidarity in the European Union
2000. ISBN 3-7908-1334-6

Johannes Bröcker/Hayo Herrmann (Eds.)
Spatial Change and Interregional Flows in the Integrating Europe
2001. ISBN 3-7908-1344-3

Kirstin Hubrich
Cointegration Analysis in a German Monetary System
2001. ISBN 3-7908-1352-4

Nico Heerink et al. (Eds.)
Economic Policy and Sustainable Land Use
2001. ISBN 3-7908-1351-6

Friedel Bolle/Michael Carlberg (Eds.)
Advances in Behavioral Economics
2001. ISBN 3-7908-1358-3

Volker Grossmann
Inequality, Economic Growth, and Technological Change
2001. ISBN 3-7908-1364-8

Thomas Riechmann
Learning in Economics
2001. ISBN 3-7908-1384-2

Miriam Beblo
Bargaining over Time Allocation
2001. ISBN 3-7908-1391-5

Peter Meusburger/Heike Jöns (Eds.)
Transformations in Hungary
2001. ISBN 3-7908-1412-1

Claus Brand
Money Stock Control and Inflation Targeting in Germany
2001. ISBN 3-7908-1393-1

Erik Lüth
Private Intergenerational Transfers and Population Aging
2001. ISBN 3-7908-1402-4

Nicole Pohl
Mobility in Space and Time
2001. ISBN 3-7908-1380-X

Pablo Coto-Millán (Ed.)
Essays on Microeconomics and Industrial Organisation
2002. ISBN 3-7908-1390-7

Ludwig Schätzl/Javier Revilla Diez (Eds.)
Technological Change and Regional Development in Europe
2002. ISBN 3-7908-1460-1

Alberto Quadrio Curzio/Marco Fortis (Eds.)
Complexity and Industrial Clusters
2002. ISBN 3-7908-1471-7

GPSR Compliance

The European Union's (EU) General Product Safety Regulation (GPSR) is a set of rules that requires consumer products to be safe and our obligations to ensure this.

If you have any concerns about our products, you can contact us on ProductSafety@springernature.com

In case Publisher is established outside the EU, the EU authorized representative is:

Springer Nature Customer Service Center GmbH
Europaplatz 3
69115 Heidelberg, Germany

Batch number: 10725973

Printed by Printforce, the Netherlands